Anti-Entropy versus Entropy

by Rolf A. F. Witzsche

Contents

5

About the Illustrated Science series
*On the Ice Age and Climate Change
and the book*

Anti-Entropy versus Entropy
Book1 of the series, Big Bang, Blow Out

The Big Bang Creation theory is a highly imaginary mythological tale that is built on the premise that 99.9999% of the universe does not exist. This means that it is so full of holes in its premise that it would make Swiss Cheese envious.

Two questions present themselves here: What is the nature of the real universe? And secondly, why is a false model intentionally created? The answers are reflected in the title. The real universe is anti-entropic. The false model is intentionally the opposite. It defines the universe as entropic. It projects the concept of entropy as a universal condition, that as a lie, excuses the economic collapse of the world, or nations, that have become subjected to systems of empire. The system of empire is anti-economic. It is entropic. Both fronts are critical to be understood, because If they are not, the Ice Age Challenge will not be addressed and be responded to, so that the coming phase shift to glaciation conditions in the 2050s will overwhelm humanity, with very few people surviving the consequences. That's why 3 books are needed to explore the issues.

Mainstream cosmology regards the universe, the galaxies, and the solar system exclusively organized by gravitational force that is known to be the weakest universal force. Mass and gravity are all that Big Bang Theory allows. However, the next higher-order force in the universe is the electric force that is 39 orders of magnitude stronger than the gravitational force. It is expressed in plasma that makes up 99.9999% of the universe. This reality is not allowed to be recognized as an organizing force in the universe, because it is expressed in electrically charged plasma that is deemed not to exist. Cosmology thereby imprisons itself with cosmo-mythologies where nothing is actually true, and humanity becomes imprisoned with it, by the false concepts. One of biggest imprisoning fudge factors, is the Big Bang theory itself.

While technology has furnished astronomy with amazing capacities for looking at the universe, ironically, what is observed is being falsely interpreted on the basis of assumptions that are simply not true, that are mythological assumptions. As a consequence, ironically, mainstream astronomy looks at the universe blindfolded. What comes out of it, of course, are tragic misperceptions. The results are often so confusing that mysterious fudge factors need to be invented to make the results appear plausible. No such fudge factors are needed in plasma cosmology.

With the next Ice Age on the near horizon, potentially beginning in the 2050s, we cannot afford to play games with fudge factors. The recognition of the true nature of the universe, the galactic system, and the solar system, that together drives the Ice Age dynamics, becomes an existentially critical issue. If humanity remains 'asleep' on this front, we may all die in the easy chair of the consequence when the glaciation conditions resume, which evidence promises, will happen quickly.

Plasma in the physical universe is as challenging in perception as the spiritual domain in the human sphere. Both are invisible, except by their effects, but they are understandable and knowable. But how does one break away from the fairy tales that inspire delusions? Answers must be found.

With the Ice Age Challenge now before us, we face two imperatives. One is to understand the real physical dynamics that power and affect the Sun, and with it to create the physical infrastructures that enable human living to continue in an Ice Age climate. The second challenge, and this is the greater challenge, is to raise up our humanity to such height as will impel us to get the job done. Some say that miracles are needed on both fronts. But what of it? Are we, as human beings, not the miracle makers on the Earth?

In the real universe, the cosmic operations are anti-entropic in nature, and expanding and progressing. We, ourselves are evidence of this progression. Should this progression have ended? Neither is our Sun isolated from the progressive nature of the universe, but expresses its dynamics, its resonating plasma streams, and their reflection in the climate on Earth. Shouldn't we develop ourselves spiritually and culturally, likewise?

Climate Change reflects the nature of the universe. It should also be reflected in us.

The Earth itself is the creation of the Sun, with its atoms having been massively synthesized in high-energy times near the center of the galaxy.

The synthesizing plasma fusion is presently at a low state, though it is currently enhanced for our Sun by electromagnetic 'Primer Fields' that focus interstellar plasma onto the Sun in a highly condensed manner. When the plasma-focusing system becomes inactive, below the required threshold conditions, the Sun reverts to a type of cosmic default level with 70% less energy being radiated, and higher rates of solar cosmic-ray flux being experienced.

At the present rate of plasma diminishment being experienced, the solar activity phase-shift threshold to the next Ice Age period may be crossed in 30 years, or in the 2050s, most likely. With the primer-fields system gone inactive by then, the climate on Earth will get 40 times colder than the Little Ice Age in the 1600s had been. Ice core evidence promises that. Without the needed preparations for human living in such an environment, 99% of humanity would die of starvation, both by the cold, and by CO_2 depletion that diminishes agriculture, as more CO_2 becomes dissolved into the sea.

With the 'Primer Fields' being critical for our very existence, the exploration of them is likewise critical.

In the Little Ice Age, between 10% and up to 30% of the populations in Europe had perished by starvation. The last Big Ice Age was evidently vastly harsher. Only 1-10 million people emerged from it alive. That's all we had after 2 million years of development. We want to do far better this time around; and we can, with large-scale technological infrastructures for our food supply. But will we create them? Will we get the job done in the 30 years that we still have left before the Ice Age starts anew? Will we even consider it? And how certain are we that the phase shift to the next glaciation period will begin, as the evidence suggests, in the 2050s? We have no slack on this front. Should we fail us on this absolute front, we would be committing suicide.

Numerous fields of evidence tell us that the next Ice Age is near. That's where the truth begins. Most of the evidence was discovered in the 1990s and thereafter. Some evidence is measured in ice cores; some is measured in space, by satellites. Some measurements are also made on the ground in terms of measurements of the Earth's magnetic-pole drift observed in northern Canada. All of this is seen combined with high-energy physics experiments at a leading national laboratory, and is also explored in the small in static experiments.

So, what will the answer be? Will we move with the evidence? Or will we lay ourselves down to die by default?

It takes an independent researcher to brake the taboos that have kept mainstream cosmology imprisoned, increasingly, during the past century, even while what is regarded as taboo is known to be wrong.

The Illustrated Science series is intended to open the scene beyond the threshold of accepted taboos, to where the actual physical evidence speaks for itself.

The scope of the existential challenge that the Ice Age brings with it, takes astrophysics out of the academic domain and places it into the foreground as one of the most-critical issues of our time. The big Climate Change events that have already worldwide effects are mere fringe effects in the flow of the ever-changing cosmic dynamics. The big effect, when the Ice Age begins anew, promises to be caused by a dimmer and colder Sun. The loss of 70% of the Sun's radiated energy defines our climate future that begins in the near term.

Sure, we can live with all that by creating new platforms for agriculture that are able to operate under Ice Age conditions. But will we do it? The task is enormous. Or will we fail ourselves on this front? We have no reason to allow us to fail. We have the materials and energy resources on hand to accomplish everything that is required for us to continue to live in an Ice Age World. But will we do it? The big question that never goes away, therefore, is; will we develop our inner resources as human beings sufficiently to get the job done, and to get it done in time? Or will we do nothing, ignore the challenge, and condemn our children and one-another to an agonizing death by starvation? That's the choice.

Towards meeting the inner challenge, I have created the epic series of novels, The Lodging for the Rose. And further, towards meeting the science challenge, I have produced numerous research books and several dozen exploration videos that the Illustrated Science series is modeled after. The work is the result of a quarter century of research, for which numerous elements of evidence in related fields came to light during the timeframe of my research.

It is my hope that the work that went into all of these projects will help in some degree - for humanity that we are all a part of - to write itself a ticket to have a future.

High-resolution color images, of the images in this book, can be obtained at www.iceagetheatre.ca

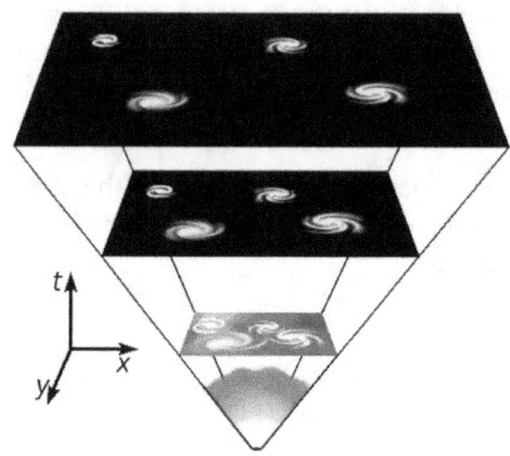

The Big Bang cosmology is built on a grand theory of the origin of the universe.

Its story begins with the postulate that nothing existed before the Big Bang 13.8 billion years ago. It is also said that whatever there might have been, collapsed into itself and sparked a giant explosion in which all the atoms in the entire universe were formed, from which, subsequently, the universe was born. It was born by condensation of the created primordial dust that is deemed to be expanding at ever-greater speed, raising away from the point of explosion.

It is said, that as the expansion 'progressed,' the speeding dust condensed into stars and planets and galaxies, which will remain active for a season, towards their eventual heat death when all the energy from the Big Bang is consumed.

The theory is built on a big mistake in perception

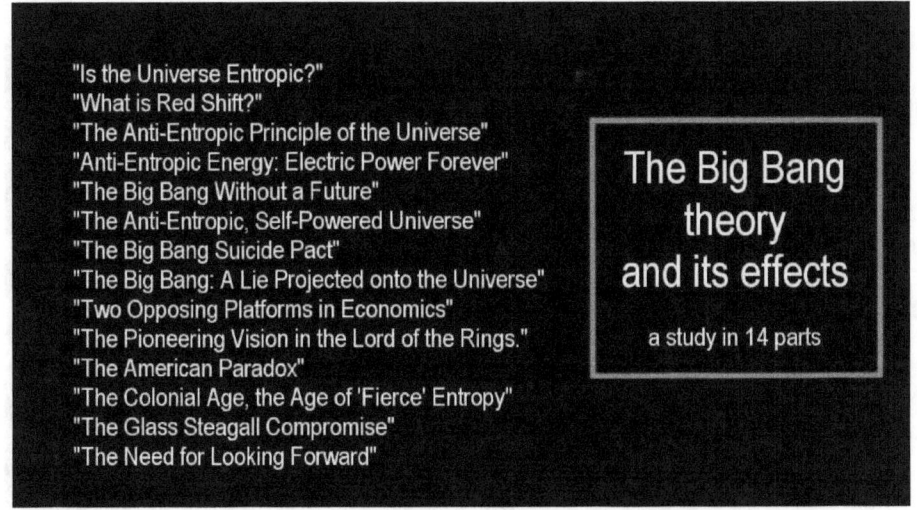

The video focuses on a big subject that calls for a big production. The subject is that big, however, it is big only because the theory is built on a big mistake in perception that has affected the shape of civilization more deeply, on a wider front, and for a longer period of time, than any other misperception in the history of humanity. The mistake in perception actually precedes the Big Bang theory by several millennia, with numerous related effects old and new that are often deemed unrelated, but which, when they are brought together into a single complex, tell a story that cannot be easily recognized otherwise. For this reason, the video became somewhat lengthy. And yes, the Big Bang theory, as a concept of the cosmos, stands in the middle of it all by its role in the larger historic context, in ways that are not apparent when the theory is looked upon, standing in isolation.

To begin, let's look at the theory itself, and the story it tells.

In the Big Bang cosmology

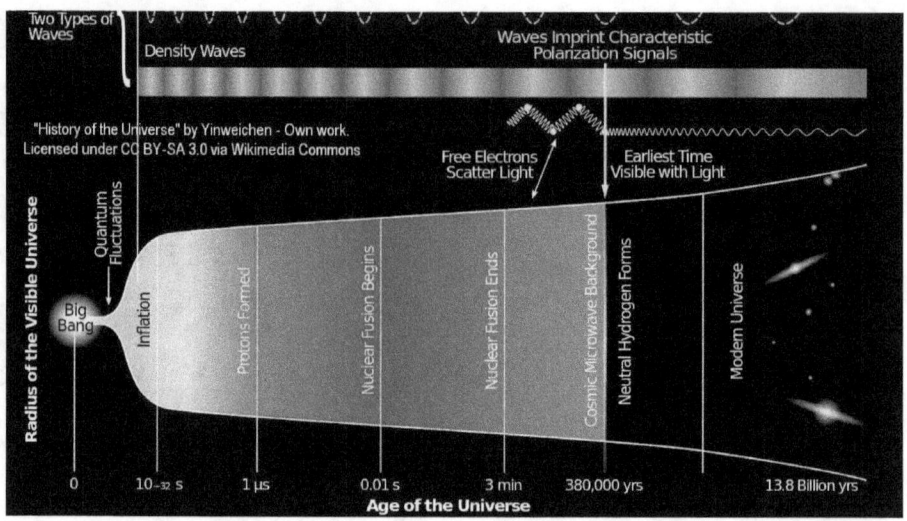

In the Big Bang cosmology, the universe is regarded to be inherently entropic in nature. All the energy and substance of the universe is deemed to have originated at a single-point source, at a single moment, as a single package. From this point on, it had all the energy and substance that it will ever have. Everything thereafter is said to have expanded outward, but with a built-in process of decay towards its death by energy depletion. The theory thereby presents the universe as entropic in nature. It is deemed to be self-consuming, winding down, and consequently self-collapsing. The term, entropy, defines an inherently collapsing dynamic system that is winding itself down by a process of depletion.

Every sun in the universe

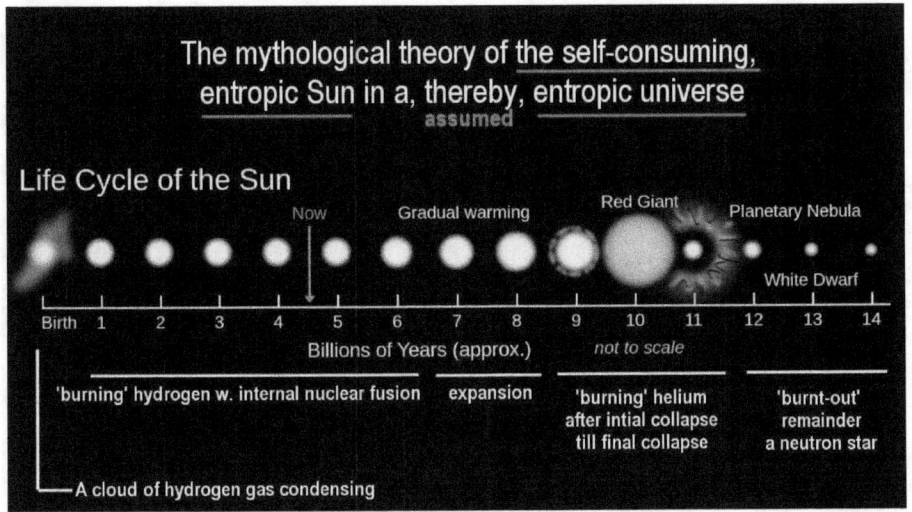

Every sun in the universe is 'seen' in this context. It is believed to be self-consuming as it burns the hydrogen gas that it is made of. The hydrogen gas is deemed to power a process of nuclear fusion deep within a sun, which is said converts hydrogen into helium with the release of free energy. By this theory, after a few dozen billion years have passed, every sun in the universe will have burned itself out, towards the end of the universe itself. The death of the universe is thereby deemed to be assured by the built-in entropy of the cosmic system that is simply unavoidable. Can this be true?

Is the Universe Entropic?

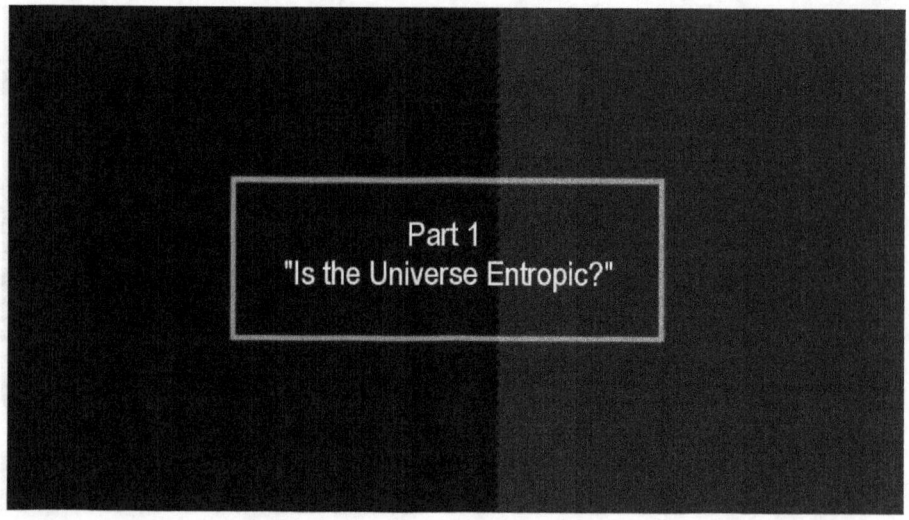

"Is the Universe Entropic?"

Entropic systems are common on Earth

Entropic Energy Systems

Attribution: I, PHGCOM   wikipedia

Entropic systems are common on Earth. We are familiar with the concept of entropy. We find entropic systems in the form of clocks, watches, and wind-up toys that run down and stop when the energy in their spring or their battery is spent. That's the effect of entropy. The Big Bang theory tells us that the universe operates in the same manner, though much more slowly. It is said to have expanded from an intense single-point explosion 13.8 billion years ago, that wound it up, that got it going, that got it expanding further and further until it fades into nothing when the energy that came with the explosion, is used up. Can this be true?

The basis for the entropic universe theory

The center of the Milky Way, at the center of the Big Bang explosion of the universe

The basis for the entropic universe theory, which the Big Bang theory may have been derived from, is the red-shift effect of light coming from distant galaxies. The more distant the observed galaxies are located, the greater is the red shift that is being observed in the light received from them. The stated theory is, that the red-shift is caused by the observed objects receding away from us. This simply means, that the greater the red-shift is, of the light from observed galaxies, the faster the galaxies are speeding away from us, the observer. But here a paradox begins to unfold that unravels the Big Bang theory.

The paradox is that the red-shift in light is observed in all directions. This means that the Earth is once again believed to be the center of the universe, as had been believed in medieval times, so that the entire universe is deemed to be speeding away from us. Wow! But is this true? Does this make sense?

In whichever direction we look, various amounts of red-shift are observed, while not the faintest blue-shift is ever observed. This evidence makes one suspicious, doesn't it?

The measured red shift is evidently real, but by it being real, it places the most fundamental platform of the Big Bang theory into doubt, and everything with it that it is built on it, because it simply makes no sense that miraculously the Earth should be the center of the universe with everything speeding away from us in all directions. The proposition places the entire theory of the entropic nature of the universe seriously into doubt.

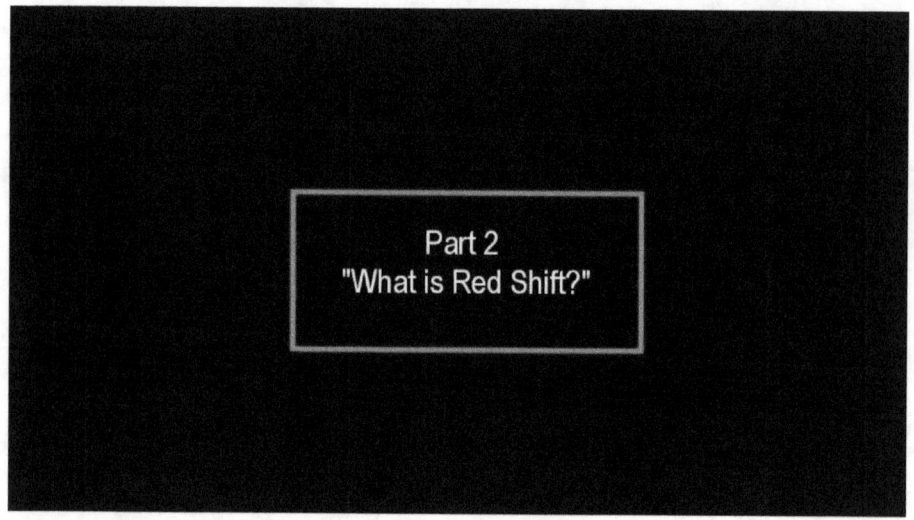

"What is Red Shift?"

When we measure red shift

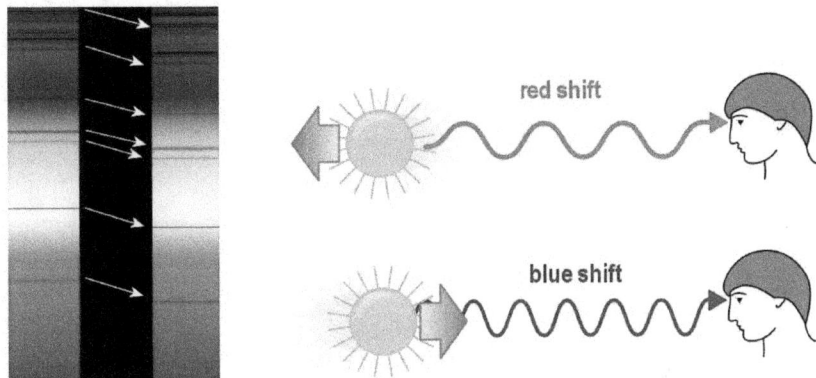

the red-shift of light from distant galaxies
is assumed to indicate that the galaxies
are fast moving away - stretching the light

What do we measure when we measure red shift?
It has been theorized that when a source of light is moving away
from an observer, the light-waves are stretched out, which causes
the observer to perceive the light shifted towards longer
wavelengths: towards the red of the visible spectrum.
Inversely, it has been theorized that when a light source is fast
approaching an observer, the light is compressed by the movement
towards shorter wavelengths, which is deemed to cause a shift to
occur, towards the blue.
The shifting itself is observed in the shifting absorption lines in the
light spectrum. The individual lines are caused by specific colors of
light being absorbed by specific atomic elements in the path of the
light. The resulting absorption pattern is known to be essentially
uniform among the galaxies, only the observed shifting of the
pattern varies. From the amount of the shifting, and from the
resulting calculated expansion speed of the most distant galaxies, it
has been theorized that the Big Bang creation event occurred 13.8

billion years ago. But is the theory true? Can it be true?

The Earth the center of the universe

The center of the Milky Way, at the center of the Big Bang explosion of the universe

Is the Earth really the center of the universe, with all galaxies racing away from us? Does the wide field of evidence that we see, match the assumption?

No, it really doesn't. The wonderful Big Bang tale appears to be full of self-evident holes. One of the biggest of these holes is the hole that doesn't exist.

The big ring of fire and the central void

A puff of smoke

ESA/Hubble & NASA
Acknowledgement: Claude Cornen   SNR 0519 - located over 150 000 light-years from Earth

When a giant explosion occurs in space, the explosion creates a thin shell of fire around what fast becomes an empty center. The result would be similar to what we see here. But this is not what we see through the telescopes, when we look at the universe. The big ring of fire and the central void that we should see, don't exist.

The entire Big Bang Creation theory

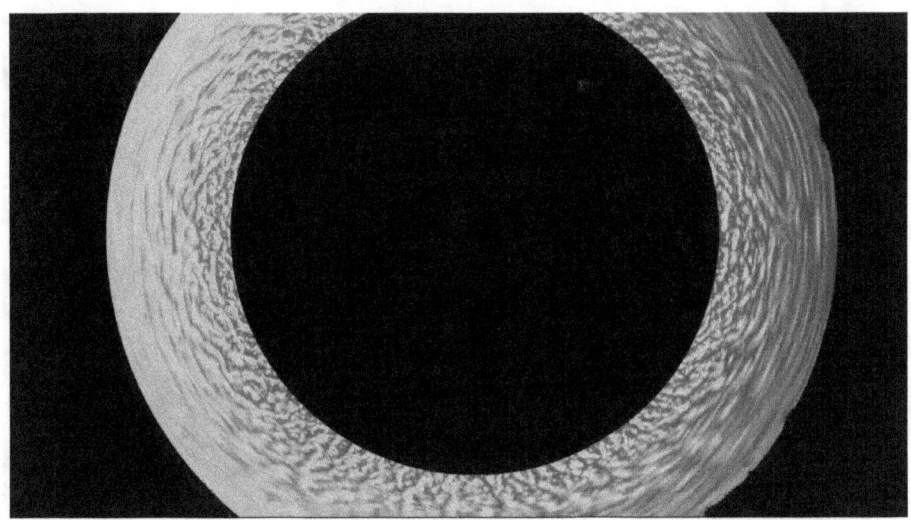

No central void has ever been found. No ring of fire with an empty center has been recognized. So what about the red-shift then? What happens to the light from distant galaxies, especially from those at the outer edge that are deemed to be racing away from us? What happens to the light that causes it to red-shift?
That's in important question, because the entire Big Bang Creation theory is fundamentally built on a specific assumption for the red shift phenomenon? How does one sort out the truth?

From the basis of plasma physics

2MASS LSS chart-NEW Nasa

The center of the Milky Way, at the center of the Big Bang explosion of the universe

When we look at the universe from the basis of plasma physics, the red-shift phenomenon, and the paradox of the missing evidence for the Big Bang, become rather easily solved. The key for this is found in the unique nature of light.

The nature of the photon, the carrier of the light

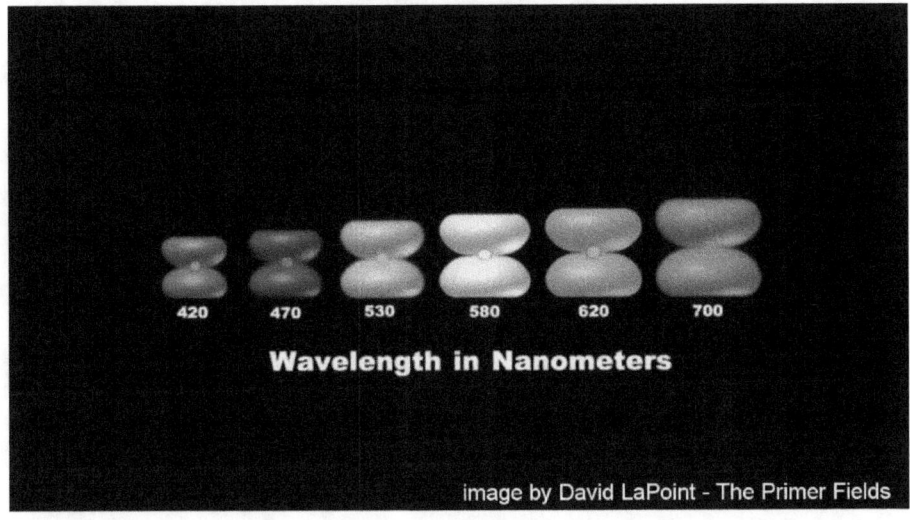

Wavelength in Nanometers

420   470   530   580   620   700

image by David LaPoint - The Primer Fields

The nature of the photon, the carrier of the light, is such that a fast moving light source doesn't actually alter the 'color' of the light. The fact is easily recognizable.

The color of light is determined by the energy environment within the electromagnetic photon envelope, when the light is created. Some photon packets are larger, some are smaller. The higher the energy level is, inside a photon package, the tighter the package is held together, and the smaller it thereby becomes. The physical movement of an atomic element in which a photon package is formed, does not affect the size of the package. Only the energy-level within the atom affects the size of the photon and thereby its color. Different size photons are recognized as different colors. This fundamental pattern, of course, extends far beyond the visible spectrum, which is actually quite narrow. Extremely high energy levels, for example, produce extremely tiny packets, such as the x-ray 'photons' that are typically 100,000 times smaller than the photons of visible light. But why won't light change its color, once it is created? Let's take a look at that.

A different principle applies to sound waves

The Tokaido Shinkansen high-speed line in Japan - Wikipedia

When one throws an apple out of the window of a fast-moving train, the apple doesn't end up being stretched out before it impacts the ground. This is so, because an apple is an entity that remains as it is, regardless of the speed of the train that carries it. However, a different principle applies to sound waves.

Light is a propagated stream of individual entities

"Gare de Lyon TGV orange" by Smiley.toerist - Own work. Licensed under CC BY-SA 3.0 via Wikimedia Commons

When a fast train drives through a station, its whistle is of a higher pitch when it approaches the station, and of a lower pitch when it moves away from the station. This is so, because a sound wave is a disturbance in the air that reaches an observer's ear at a different rate by approaching or departing. In sound, the shifting pitch is normal, because sound is a disturbance in the air, while light is not a disturbance, but is an individual object, a discrete package. A beam of light is a propagated stream of individual entities.

Light consists of photons that are individual entities

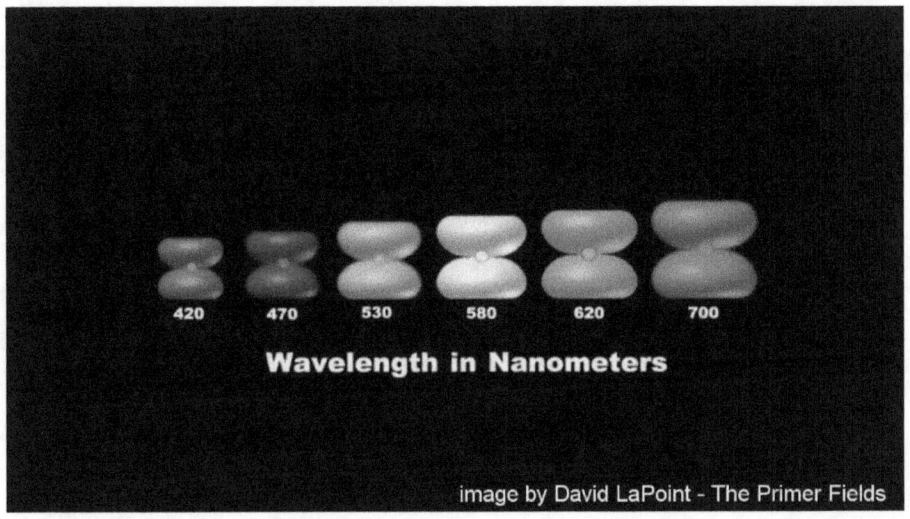

image by David LaPoint - The Primer Fields

However, because light consists of photons that are complete individual entities that are of a specific size for a specific color, and the size of the photon is determined by the energy it contains, the size can change when its energy is dissipated on the path of its travel over extremely long distances.

How is the red-shift possible?

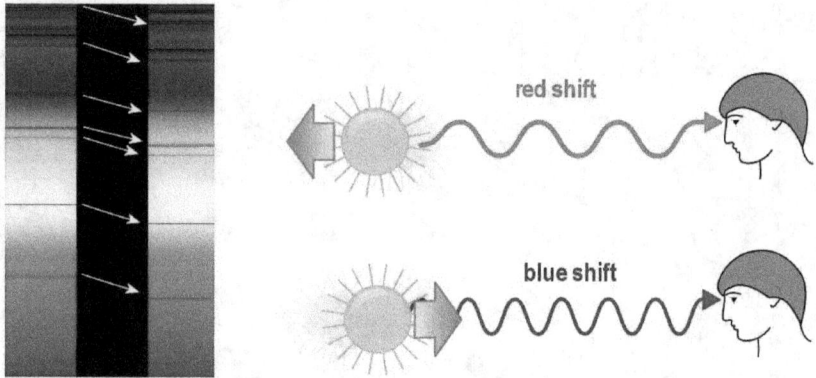

red shift

blue shift

the red-shift of light from distant galaxies is assumed to indicate that the galaxies are fast moving away - stretching the light

This factor is significant in the context of the Big Bang red-shift theory. The red shift that is observed in light from distant galaxies is, of course, totally real. We see the spectral lines received from distant galaxies. We see them shifted collectively towards the red. But how is the red-shift possible when light is made up of discrete photon entities that cannot be stretched in the same manner as a disturbance would be stretched by a receding source?

A slight energy depletion occurs along the way

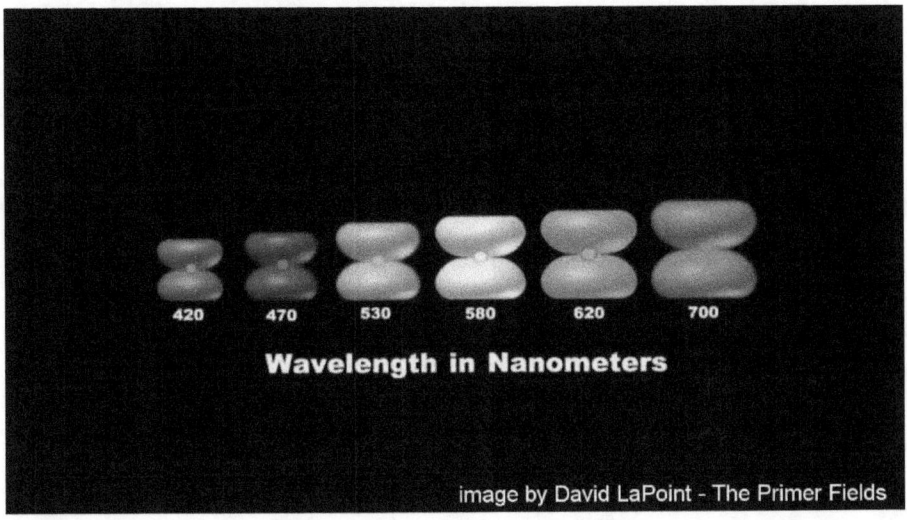

Wavelength in Nanometers

420 470 530 580 620 700

image by David LaPoint - The Primer Fields

The answer is simple. When light is propagated over very long distances in the range of hundreds of millions of light years, a slight energy depletion occurs along the way. The photon package increases in size by the energy depletion. The purple light thereby becomes blue light, it becomes a larger package, and the blue light becomes green light, and the green light becomes yellow light, and so on. The entire spectrum of the light becomes shifted towards the red, and the red, of course, gets shifted off the visible spectrum. Now, the red shift phenomenon makes sense. It has nothing to do with light from distant galaxies being stretched by a moving light source that is racing away from us. The entire Big Bang theory, thereby falls apart.

Yes, the wavelength of light appears to be stretched thereby, but this is the result of the changing size of photons that results from gradual energy depletion.

The amount of the red-shift is typically a factor of the distance that light has to cross from far-away galaxies before it arrives at our door.

Red-shift amounts do vary with local conditions

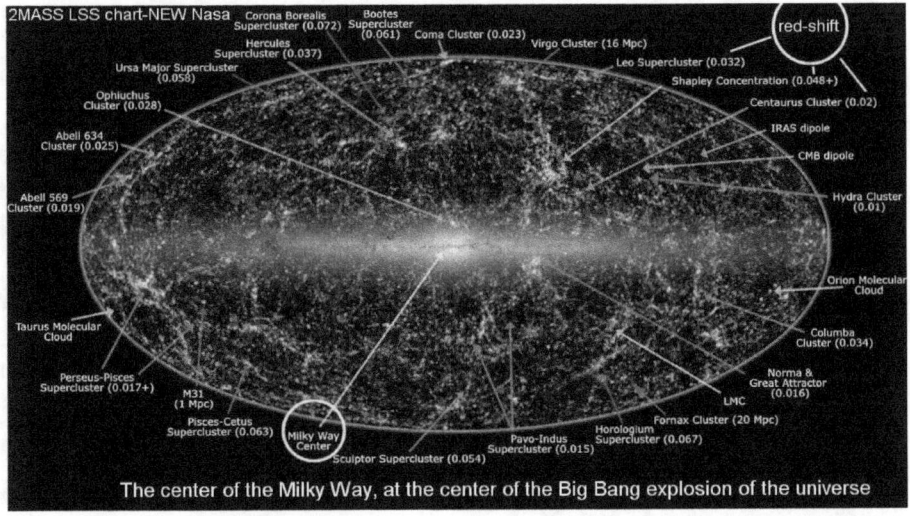

The center of the Milky Way, at the center of the Big Bang explosion of the universe

The amount of the red-shift is also a factor of the density of the cosmic dust, gases, and plasma that the light encounters along the way. This too, affects the rate of energy depletion.

For this reason, the red-shift amounts do vary with local conditions, both at the source, and along the way. The red shift definitely is not an indication that galaxies are racing away from us in all directions, as if we were at the center of the Big Bang universe.

The Big Bang theory is paradoxical

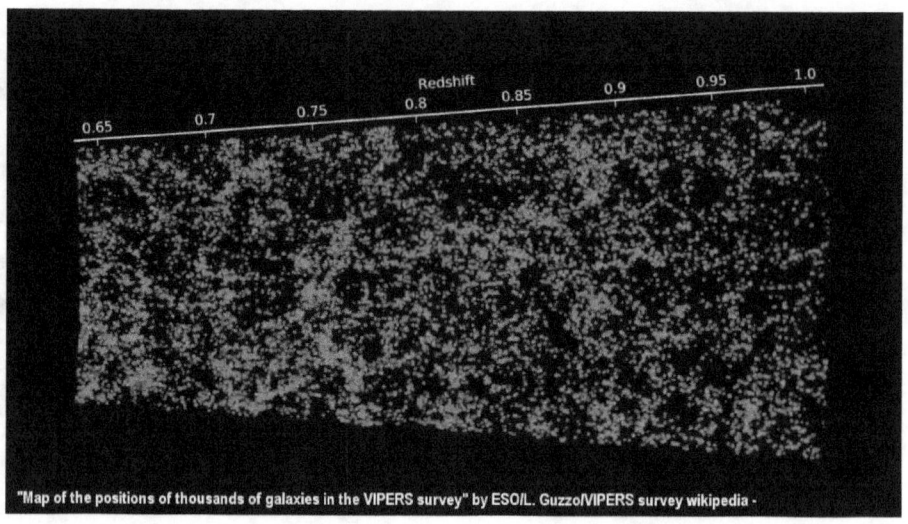

"Map of the positions of thousands of galaxies in the VIPERS survey" by ESO/L. Guzzo/VIPERS survey wikipedia -

The Big Bang theory is paradoxical for also another reason. This too, is rather obvious. The theory states that all matter in the universe was created in one place in the first 3 minutes of the explosion, and has been expanding outwardly thereafter for 13.8 billion years. If this was true, the material density of the universe would diminish outward with the cube of the distance from the source. At the rate of expansion that is theorized, the material distribution in distant regions should be so thinly spread that almost nothing should exist in distant places. Instead, the opposite is true.

"The Anti-Entropic Principle of the Universe"

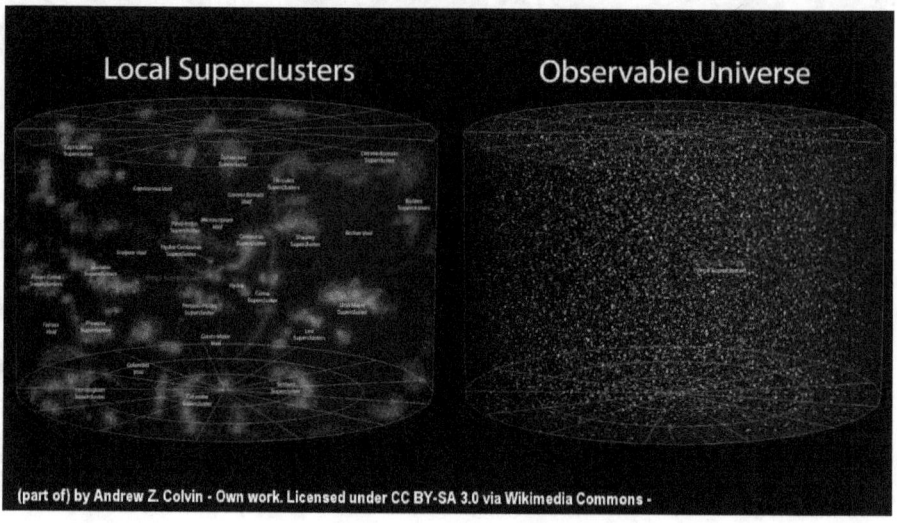

(part of) by Andrew Z. Colvin - Own work. Licensed under CC BY-SA 3.0 via Wikimedia Commons -

We see a universe that is of near homogenous density in populations of galaxies, clusters of galaxies, and super clusters. This evenly packed universe that we see was evidently not created from a single source at a single time, but is a universe that manifests itself as a self-unfolding plasma universe that is formed by creative principles that are manifest everywhere throughout cosmic space. The evidence suggests that the universe is not entropic in nature and winding down from its initial infusion of energy, but is completely self-creating everywhere, unfolding itself in a process that is opposite in nature to entropy. This opposite of entropy, one might term, anti-entropy.

David Bohm

The successor

Albert Einstein (1879-1955)   David Bohm (1917-1992)

David Bohm, whom Einstein is said to have called his successor, speaks of cosmic space as latent energy that has an implicate order and an explicate order.

The modern perception of sub-atomic particles

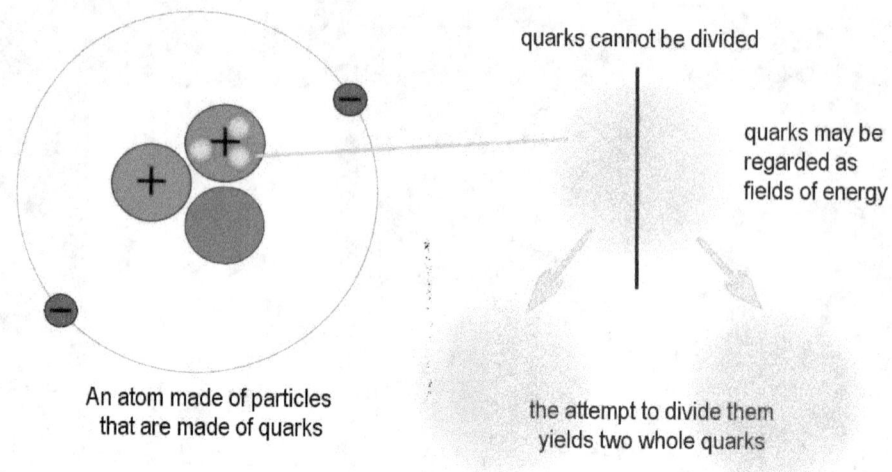

quarks cannot be divided

quarks may be regarded as fields of energy

An atom made of particles that are made of quarks

the attempt to divide them yields two whole quarks

With the modern perception of sub-atomic particles, such as quarks, as being made up of moving points of energy, which subsequently make up the protons and electrons of the universe, it can be said that everything in the universe has ultimately been derived from the latent energy of space itself. Thus, the energizing source for the universe is present everywhere throughout all cosmic space, from which everything that has become visible is derived, and also the plasma particles that are not visible.

That's the mark of anti-entropy.

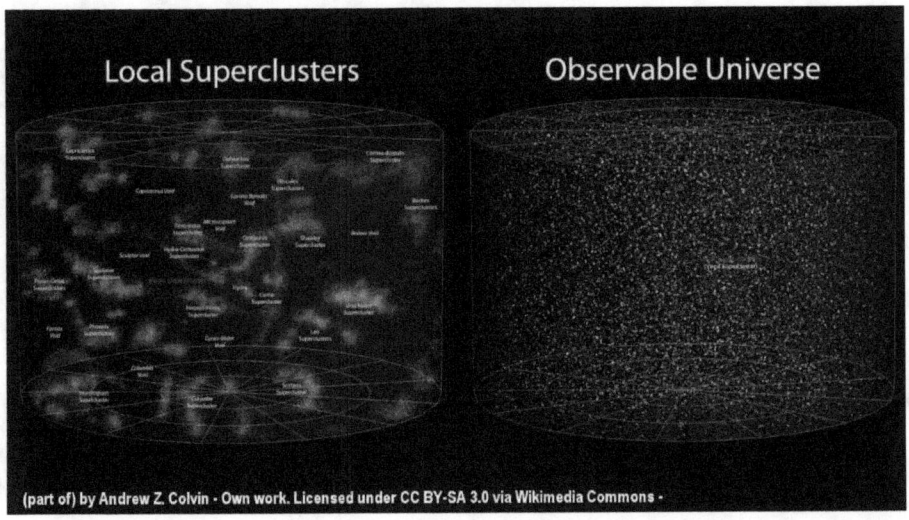

With the electromagnetic principles that set the resulting plasma into motion, stars and galaxies of stars are formed everywhere in space, which, by means of plasma fusion, synthesize all the atomic elements that exist in the universe in a creative process that happens everywhere and never ends, by which the universe grows and becomes more massive. That's the mark of anti-entropy.

Every sun a creator of its surrounding worlds

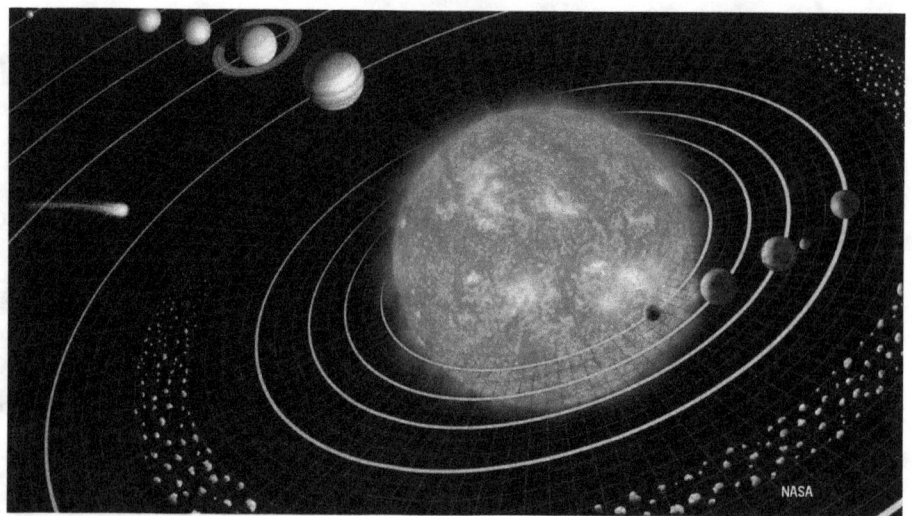

Every sun becomes thereby a creator of its own surrounding worlds. Everything that we see, every atom in the universe, has been created by a sun. This means that a sun that is not an entropic hydrogen-fusion furnace that consumes itself as the Big Bang theory would have it to be, such as a sun condensed from primordial gases - but is instead a plasma sun that is itself the creator of all the hydrogen atoms and other atomic elements that surround the sun and form the solar system. Our Sun, according to all evidence, is a plasma sun and is powered by electric plasma interaction.

The so-called 'Pillars of Creation'

"The Pillars of Creation"
Part of the Eagle Nebula

NASA,
Jeff Hester, and
Paul Scowen
(Arizona State University)

The so-called 'Pillars of Creation' of the Eagle nebula, are pillars of atomic dust synthesized by a powerful sun located behind each one of the pillars near the top. The Big Bang theory states that the dust of the pillars have created their respective sun as a product of accretion. In the real universe the opposite is true. The evidence is rather simple, that it was the sun behind each pillar that has created the atomic elements that are visible here as clouds of dust and gases. The so-called Pillars of Creation, are pillars of smoke and dust that have been created by an extremely active sun. The sun, thereby, becomes is creative source, rather than the created object, as the Big Bang theory would have it to be.

Every sun fuses plasma into atoms

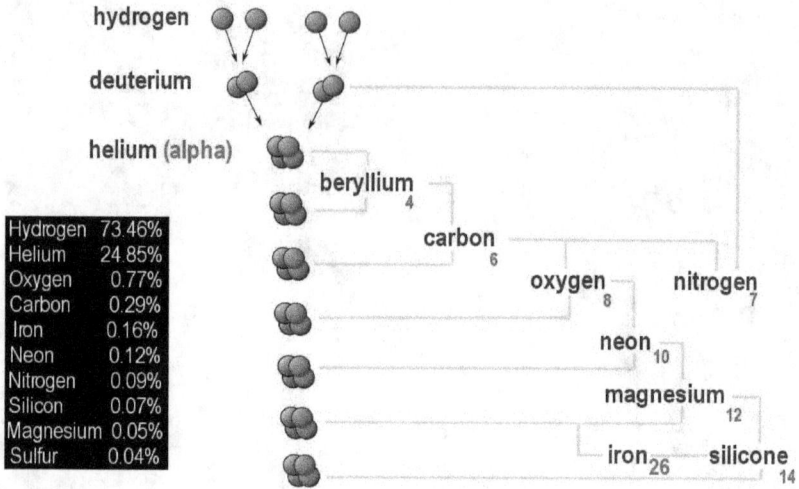

Hydrogen	73.46%
Helium	24.85%
Oxygen	0.77%
Carbon	0.29%
Iron	0.16%
Neon	0.12%
Nitrogen	0.09%
Silicon	0.07%
Magnesium	0.05%
Sulfur	0.04%

Every sun fuses plasma into atoms. It synthesizes every atomic element that exists. No Big Bang explosion is needed for a universe to be blessed with an abundance of elements. The real universe is self-creating, self-powering, and is self-maintaining and self-advancing, by the dynamics of its timeless principles that have no beginning and end.

Anti-Entropic Energy

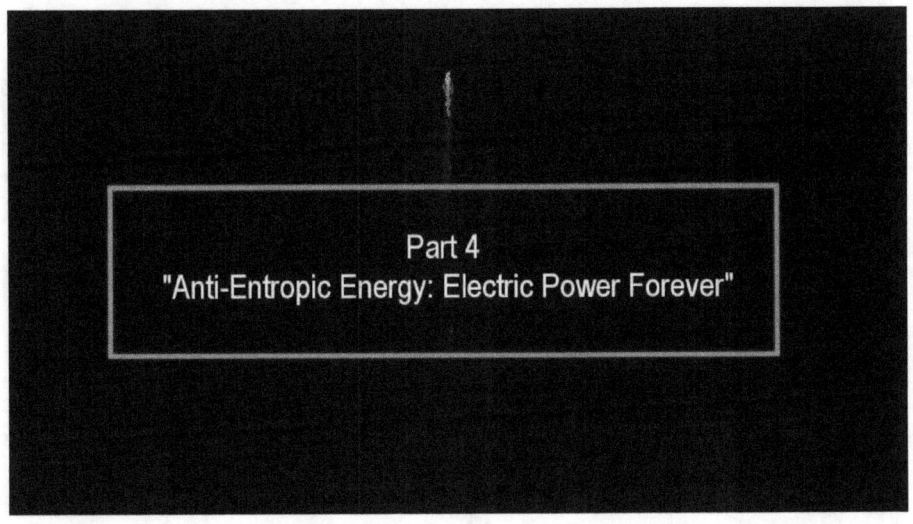

"Anti-Entropic Energy: Electric Power Forever"

Celebrate that the Big Bang theory is false

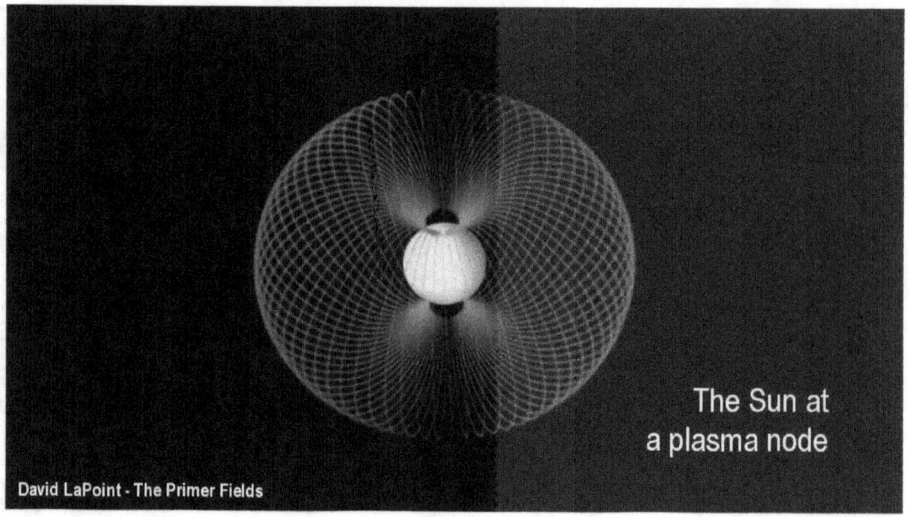

The Sun at
a plasma node

David LaPoint - The Primer Fields

We should celebrate that the Big Bang theory is false. The celebration frees us to celebrate the universe as it is. It enables the recognition that the universe is powered by immensely large, though invisible, cosmic streams of plasma that power every sun, which we too, can access for utilization on the Earth.

Not a single sun in the universe is self-powered

NASA - view from ISS

Not a single sun in the universe is self-powered. Every sun is powered by the universe directly.

The universe is powered on the cosmic scale

David LaPoint - The Primer Fields

The universe is powered by means of electromagnetic principles on the cosmic scale, that focus electric cosmic plasma streams into a sun.

Our future depends on this utilization

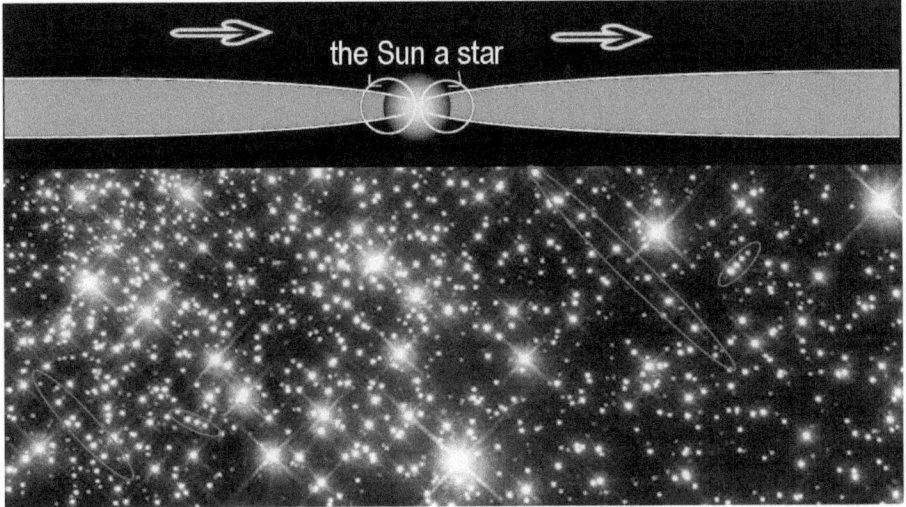

We should celebrate this truth, because the galactic electric energy streams present to us an anti-entropic electric energy resource to power our world with, as we make it available to us. Our future depends absolutely, on this utilization.

Burning fuels for energy production

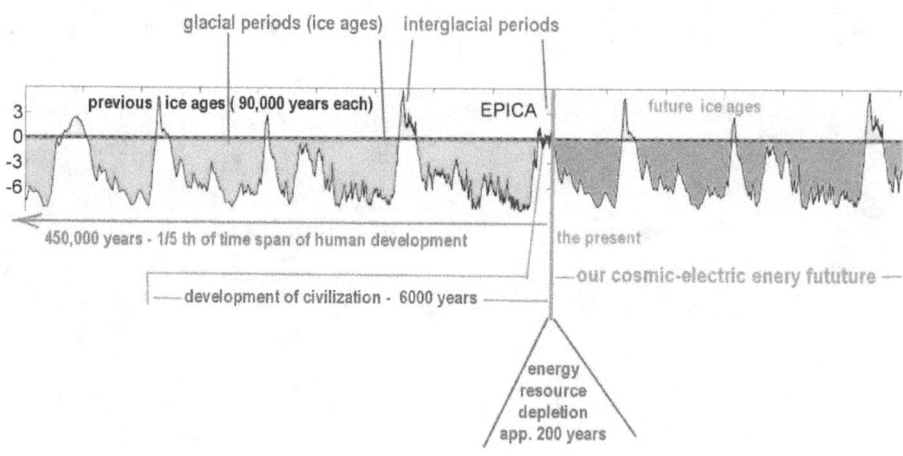

We presently live by burning fuels for energy production. These fuels are being depleted. Oil and gas and nuclear fission fuel may be depleted in 60 years. Coal may last a bid longer. Most of the depletion has occurred in the last 200 years in the 6000 years' history of our civilization. Before large-scale energy development began, the only energy resource we had, was wood taken from the land, or oil from fish. The poverty that corresponds with this primitive living, has kept humanity small.

However, once we freed ourselves from the primitive poverty, we found that the needed energy resources that give us our advanced freedom from poverty, are inherently finite. Thus, the question needs to be asked, "what will happen to us when the resources that we rely on have been used up. We may be at this stage in 100 years, or the final depletion may be delayed a 1000 years if we curtail energy use with murderous consequences. When we get to this point, will we lay ourselves down to die? Even if we would restart the nuclear fast-breeder technology that enables the fuller utilization of the available nuclear fuel, by a factor of 20, we would still run out of fuel in roughly 1000 years?

The bottom line is, that our terrestrial fuel resources are so minuscule and finite that we won't have anything left of them 1000 years from now, or potentially much sooner than that.

What will power our economy then, through the next 90,000 years of the coming glaciation period, which will be upon us, potentially, in the 2050s? What will power our world when the presently used energy fuels are depleted? What will power our world for the many millions of years into the future that humanity has the potential to have on the Earth?

To tap into the cosmic electric energy streams

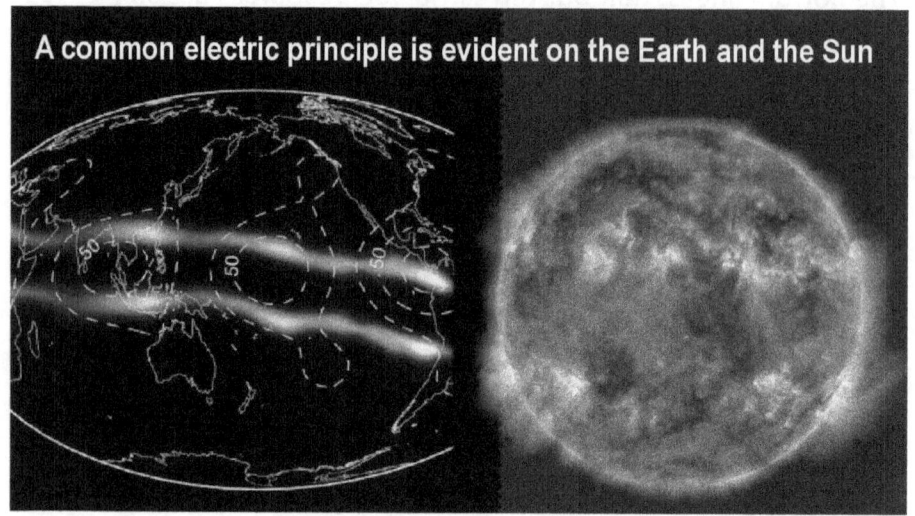

A common electric principle is evident on the Earth and the Sun

If we didn't have the option to tap into the cosmic electric energy streams that power our Sun, which are evidently available also on the Earth, our future would end at the time we run out of fuels. We would simply die back to minuscule numbers or become extinct. Fortunately, this tragedy does not need to be our future, because the Big Bang theory that blocks our vision of the real universe, is not true.

The cosmic electric energy streams that power every sun, including our Sun, do exist. NASA has even photographed them surrounding the Earth.

Plasma streams that are encircling the Earth

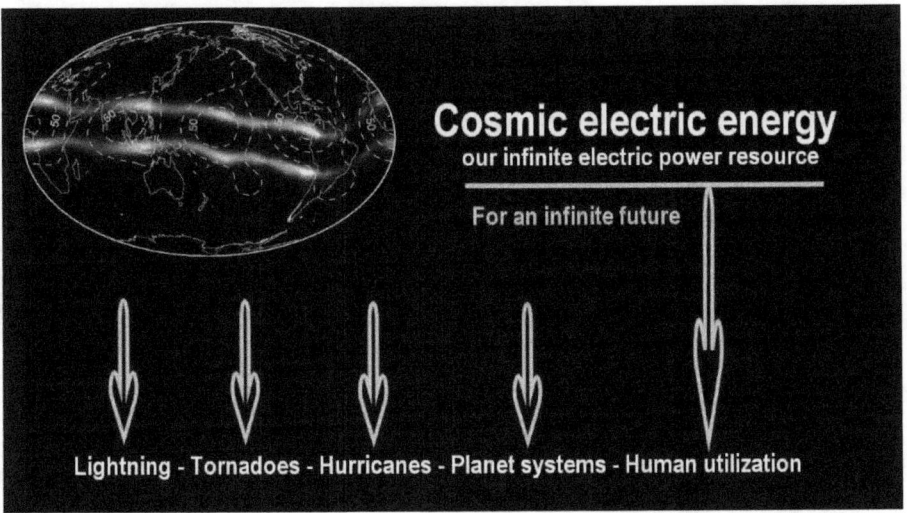

The cosmic plasma streams that are encircling the Earth in the form of two bands centered on the magnetic equator, are cosmic energy streams that promise to be available to us for our future. Of course, the technology for the utilization of this resource needs yet to be developed. And this is only a technological step away. The Big Bang theory, in contrast, would have us believe that a cosmic energy resource outside the bounds of the Earth doesn't exist. Fortunately, the theory is wrong. It has no foundation. Let's celebrate that the truth is much grander, which gives us an infinite future.

The Big Bang Without a Future

"The Big Bang Without a Future"

The Big Bang theory insists

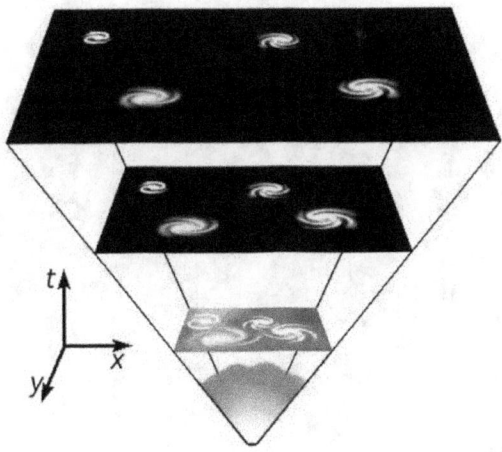

The Big Bang theory insists that humanity has no future; that the universe itself has no future. It insists that the universe was given one single shot of energy, one single infusion, and that's all it got, which implies that our tiny portion on the Earth is the totality that we will ever have.

When the energy resources are gone, our future ends

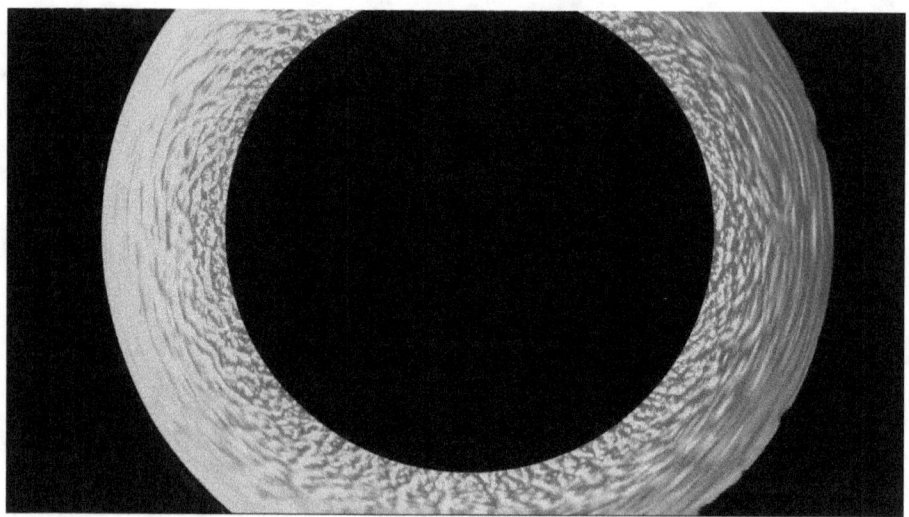

The theory says that energy was spread across the universe by the big explosion, so that what we got from it, is all we'll ever have, so that by our burning it, our future is diminishing. In other words: when the energy resources are gone, our future ends.

The concept by which everything ends

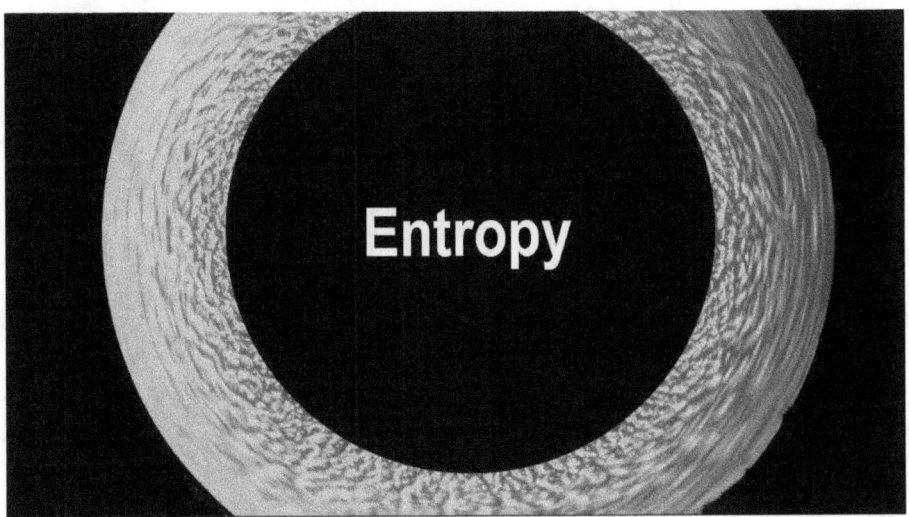

The concept by which everything ends, is termed, entropy. The Big Bang theory is entropic, because it says, that once we have used up what we were given, we will have nothing and we will die.

Even the Sun will die

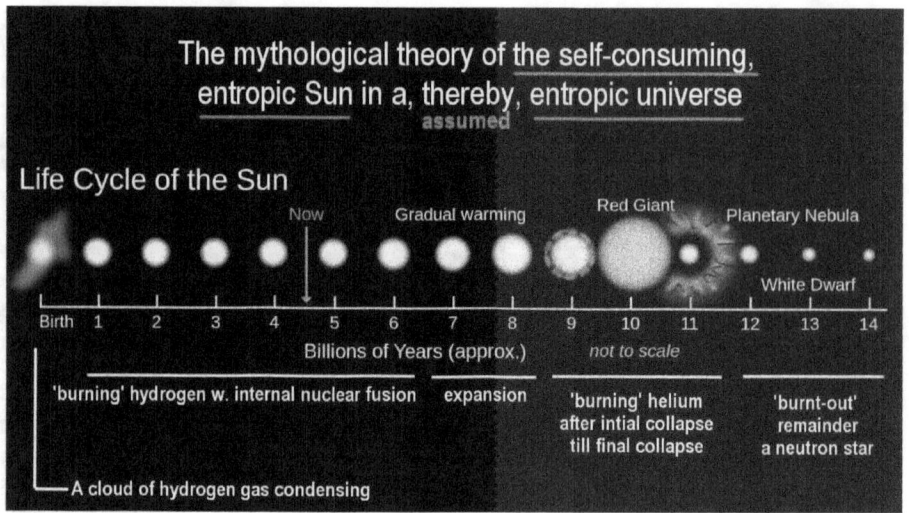

The Big Bang theory says that even the Sun will die when it used up its fuel. It says that the entire universe will grind down to nothing and die the inevitable energy-depletion death.

Fortunately, we see no evidence

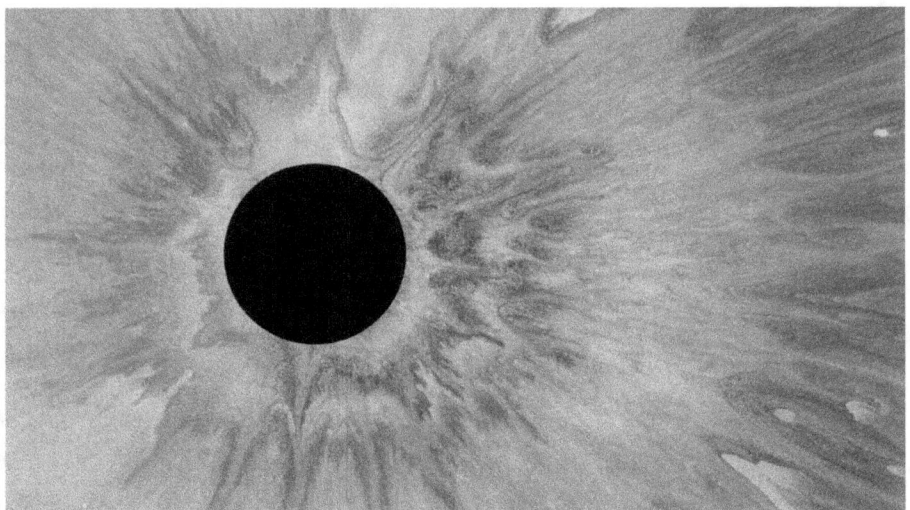

Fortunately, we see no evidence that the theory is even remotely credible. If a big explosion had created all matter in the universe, which the explosion expanded and forged with it all visible forms, then the distribution of matter would have become diffused with the cube of the distance, according to the dynamics of an explosion. This means that the primordial matter would have become spread so thin farther out that almost nothing would be found in distant places, much less the gigantic volumes of it that supposedly condense into clusters of galaxies. If the Big Bang theory was correct, there shouldn't be anything much to be seen in the universe. Of course, this is not how the universe reveals itself to the viewers with telescopes.

The observable universe is uniformly dense

2MASS LSS chart-NEW Nasa

The center of the Milky Way, at the center of the Big Bang explosion of the universe

The observable universe is uniformly dense with near even distribution of galaxy super-clusters throughout the vast reaches of space that telescopes can observe, ranging from the local super clusters to the vast sea of super clusters that are spread out across the infinite realms of the cosmos.

We are seeing a universe that is actively self-powered

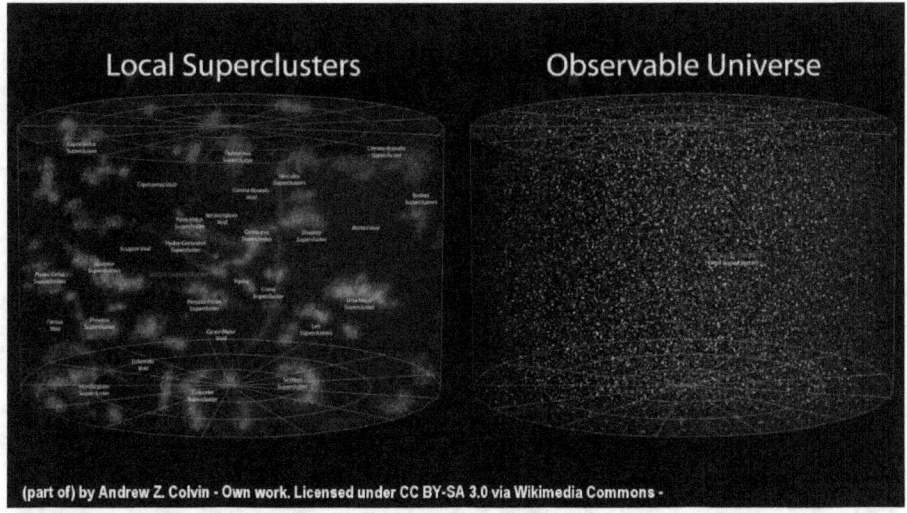

Local Superclusters   Observable Universe

(part of) by Andrew Z. Colvin - Own work. Licensed under CC BY-SA 3.0 via Wikimedia Commons -

We are looking at a universe with our telescopes, that is evidently self-created everywhere, instead of having originated from a single-point source. We are seeing a universe that is actively self-powered, and is actively self-maintained, and this everywhere with nothing running down.
But how is this possible? How can an entire universe be self-powered?

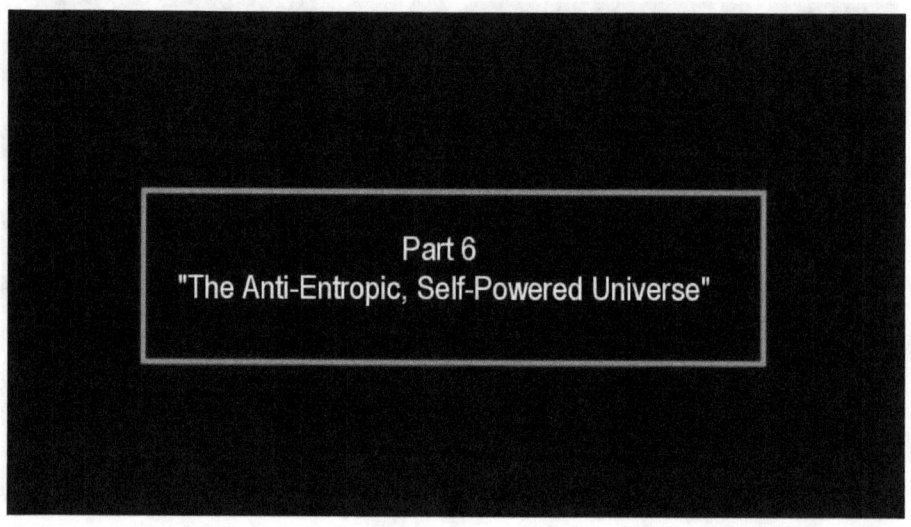

"The Anti-Entropic, Self-Powered Universe"

Back to David Bohm

The Implicate Order and Explicate Order

David Bohm
1917 - 1992

This subject takes us back to David Bohm, whom Einstein is said to have called his successor. As I said earlier, David Bohm speaks of cosmic space as not being empty, but as being a vast sea of latent energy with an implicate order and an explicate order. But what does this mean?

A latent sea of water

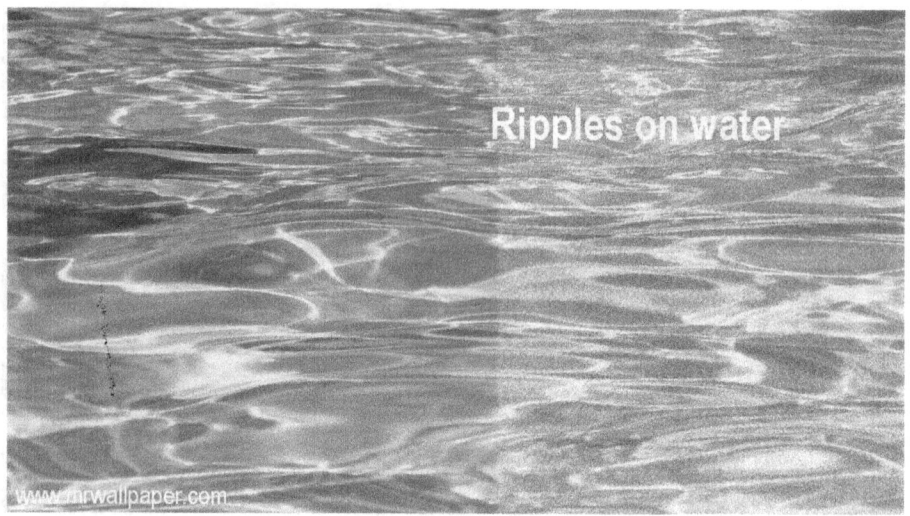

Our oceans can be seen as a latent sea of water. Nothing much happens there. However, a tiny portion of the water constantly evaporates at the surface.

Water vapor

The water vapor, though, is too small in size to be visible by itself. Only when water molecules latch together and form tiny droplets, will the evaporated vapor become faintly visible as fog.

The latent background of energy

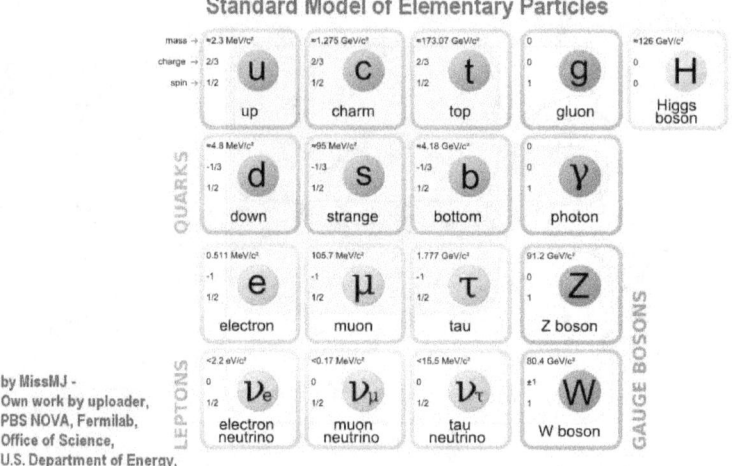

Standard Model of Elementary Particles

The latent background of energy that David Bohm speaks about, may be likened in comparison to a vast ocean of energy, from which tiny parts become discrete, become explicate, like molecules of water vapor that become discrete from the oceans. In space, the discrete explicates of the latent energy become the basic elementary particles of the universe. The major groups of these explicate discrete entities of energy, are termed quarks and leptons. The quarks have built-in characteristics that combines into specific forms, that form the proton particle, and in a similar manner the leptons for the electron particle.

The electron

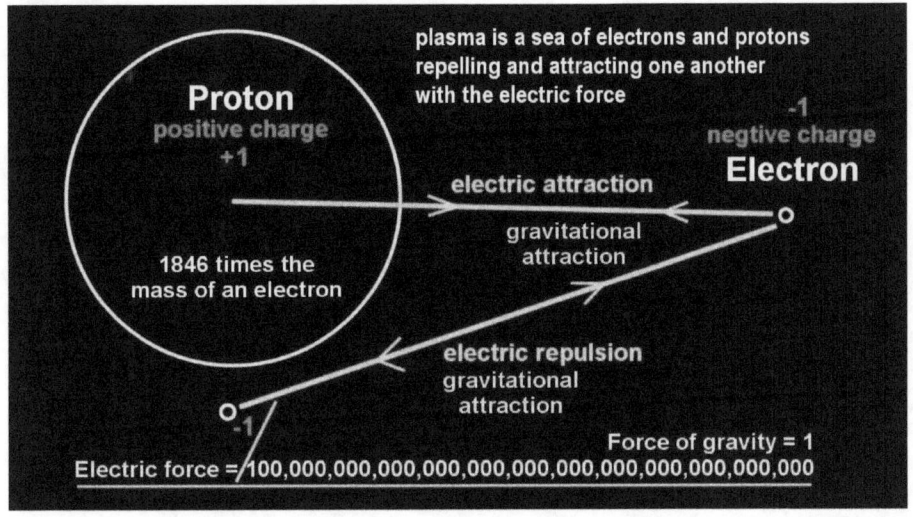

The electron is the smallest of the particles. It is a thousand times smaller than a proton.

A proton is made up of three quarks

The quark structure of the proton

The proton is made up of 3 quarks:
2 up quarks (2/3 charge)
1 down quark (-1/3 charge)
Held together by the strong force
(its total charge = 1)
Its mass is dynamic (80 to 100 times the rest mass of the quarks
The strong force is mediated by gluons (wavey)

A proton is made up of three quarks that are bunched together, which gives the proton a substantial size, and a substantial apparent mass.

The electron, in comparison

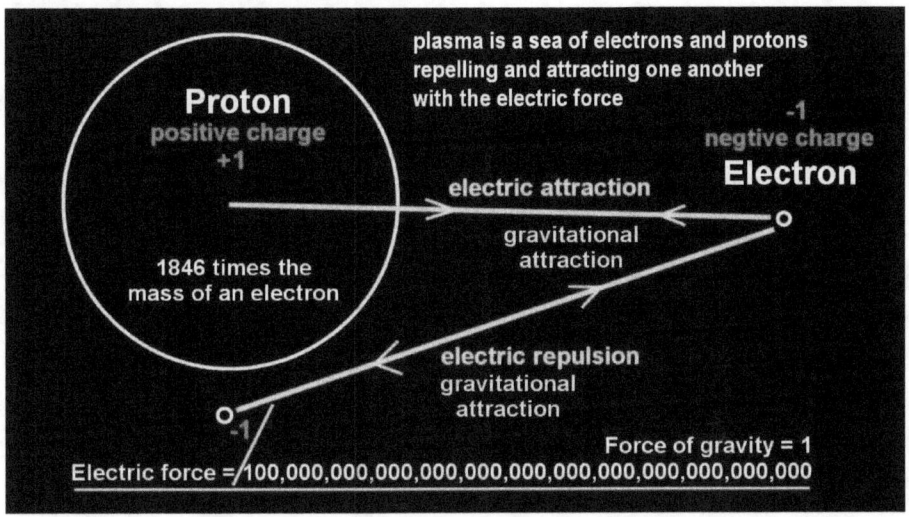

The electron, in comparison, is so small that it exists partly as but an energy wave, and only partly as a particle.

Quarks and leptons are energy in motion

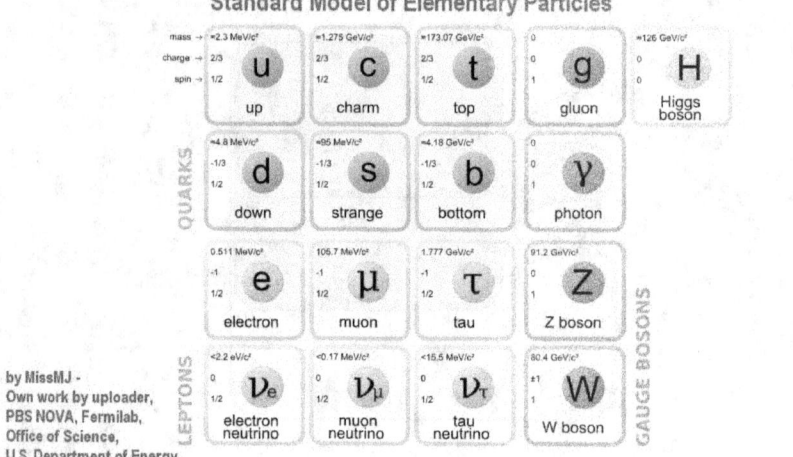

This means that the quarks and leptons, which are but energy in motion, are the basic substance for the basic building blocks of the universe.

Protons and electrons exist in space

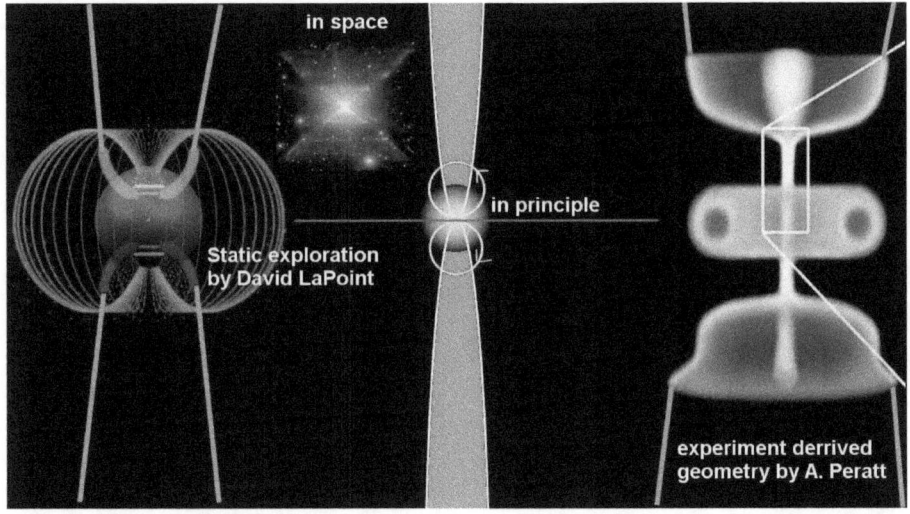

The protons and electrons exist in space, organized into vast streams termed plasma. The plasma in turn carries the forces by which the visible universe is formed and organized, and gains its mass.

Roughly 99.94% of the mass of the universe

The quark structure of the proton

The proton is made up of 3 quarks:
2 up quarks (2/3 charge)
1 down quark (-1/3 charge)
Held together by the strong force
(its total charge = 1)
Its mass is dynamic (80 to 100 times the rest mass of the quarks
The strong force is mediated by gluons (wavey)

"Quark structure proton" by Arpad Horvath
- Own work. Licensed under CC BY-SA 2.5
via Wikimedia Commons -

Since roughly 99.94% of the mass of the universe is provided by the protons, and the protons are made of quarks that are theorized entities of energy, it may be useful to look at the protons once more, and the quarks that form them. The quarks clump together in groups of three. Each carries a specific electric charge, which, when they are grouped together, give the resulting proton a positive electric charge, which in turn complements precisely an electrons negative charge. Without this precise balance, which is essential in constructing the visible universe that is made up of atomic elements, the visible universe and its higher-order constructs, with intelligent life at its pinnacle, would not exist.

Plasma, the lifeblood of the universe

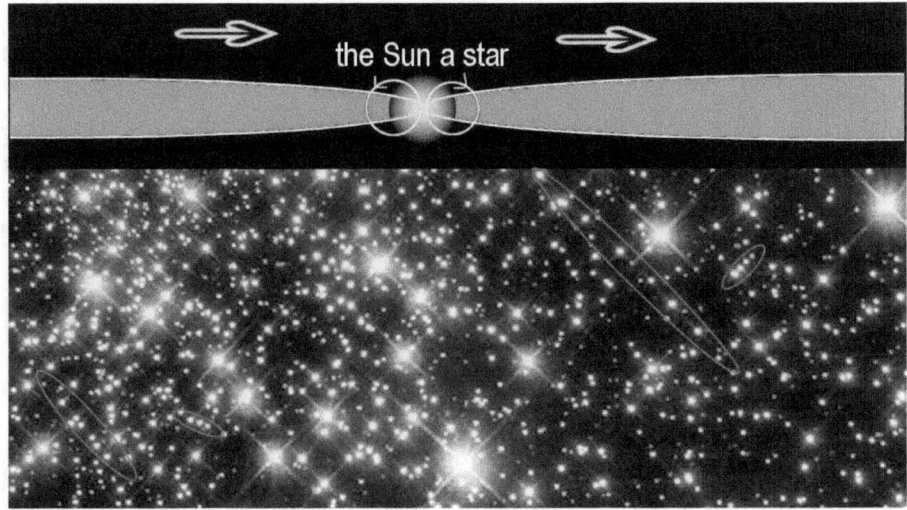

This means that the substance of the universe in all its forms is energy that exists simply everywhere, and is constantly created everywhere. Thus, the pure energy entities that unfold as quarks and leptons, are through the constructs of plasma, the lifeblood of the universe.

The universe without plasma is inconceivable

This 'blood' of energy flows everywhere, from which everything is created. The universe without plasma is inconceivable, except in dreams. It renders the universe as anti-entropic in nature.

All space is filled with plasma

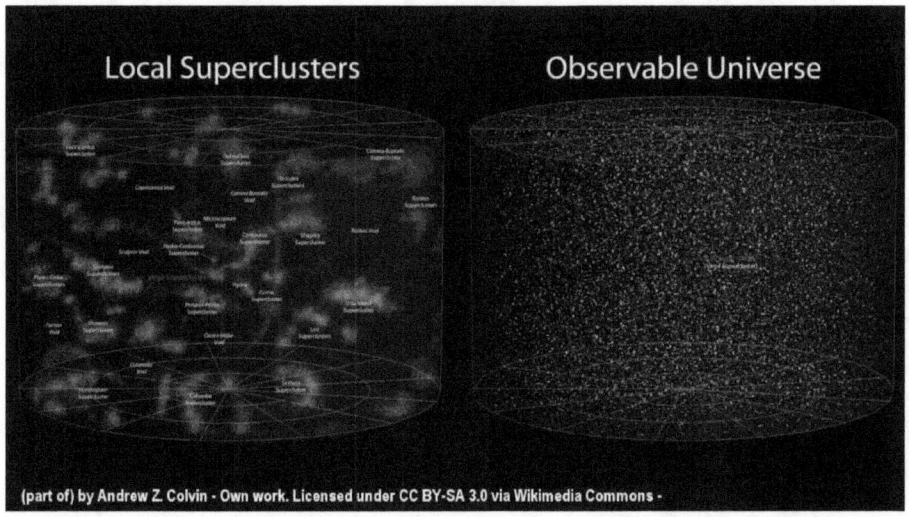

Local Superclusters

Observable Universe

(part of) by Andrew Z. Colvin - Own work. Licensed under CC BY-SA 3.0 via Wikimedia Commons -

All space is filled with plasma. It combines into giant streams. The streams of plasma have formed every sun, and continue to do so. Plasma also carries the energy that powers every sun and by which plasma becomes fused in high energy processes on the surface of a sun in an electric synthesis where the atoms of the universe are forged. This process happens now, and always had, and always will.

Every atom that exists

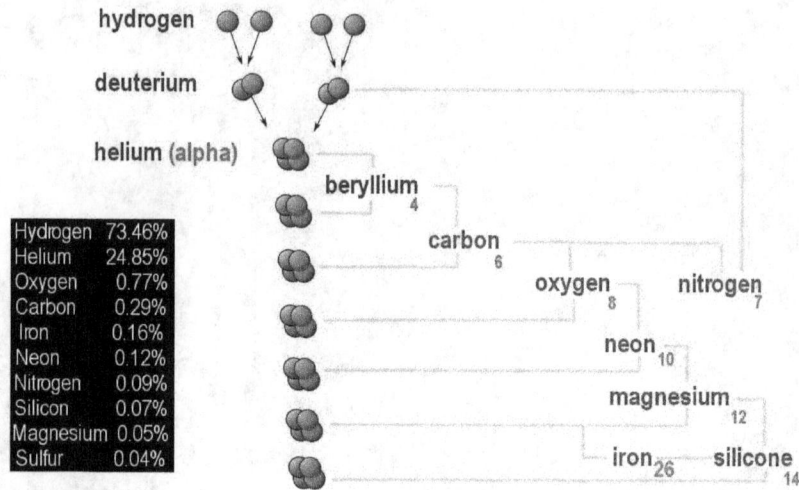

hydrogen

deuterium

helium (alpha)

beryllium 4

carbon 6

oxygen 8   nitrogen 7

neon 10

magnesium 12

iron 26   silicone 14

Hydrogen	73.46%
Helium	24.85%
Oxygen	0.77%
Carbon	0.29%
Iron	0.16%
Neon	0.12%
Nitrogen	0.09%
Silicon	0.07%
Magnesium	0.05%
Sulfur	0.04%

Every atom that exists in the universe was synthesized from plasma in electric fusion on a solar surface.

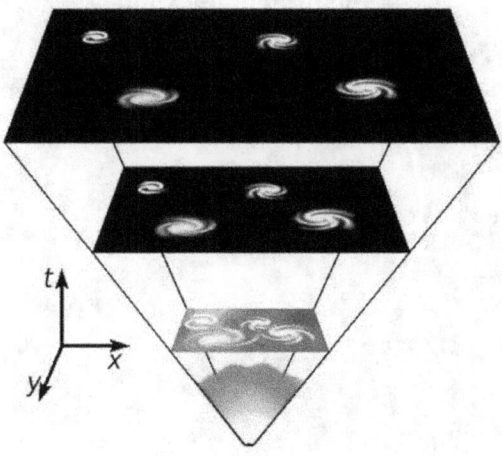

The Big Bang creation theory is nothing more than just a dream.

The creative process

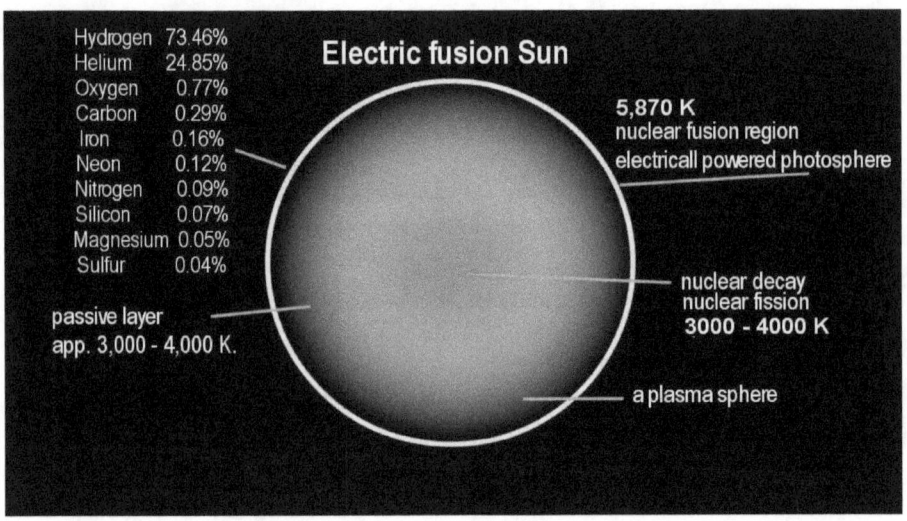

The creative process, of course, continues. It continues everywhere in the universe simultaneously. It is anti-entropic.

Nothing ever had a single-point origin

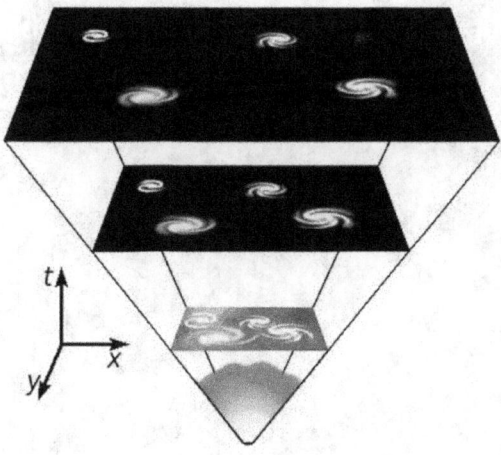

Nothing ever had a single-point origin, and collapse as its destiny.

The creative process is universal

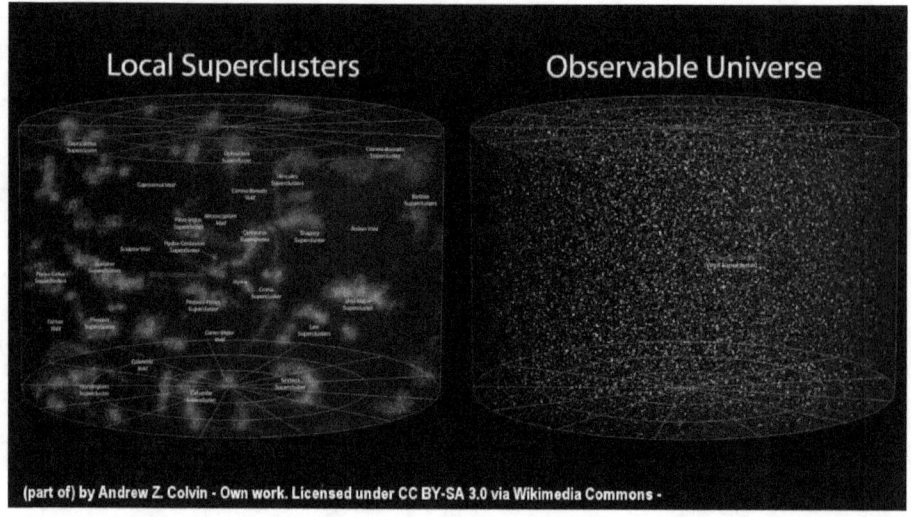

The creative process is universal. What you see here is what an anti-entropic universe necessarily looks like, and does look like.

Nuclear-fusion energy production

Joint European Torus

All this also means that the fabled nuclear-fusion energy production that is pursued on Earth, is not a viable option. The theorized basis for nuclear-fusion power is wrong, because the Sun, that it aims to replicate, is not inherently an energy producer, but acts as an energy converter of the energy that flows in the universe.

A sun is not entropic in nature

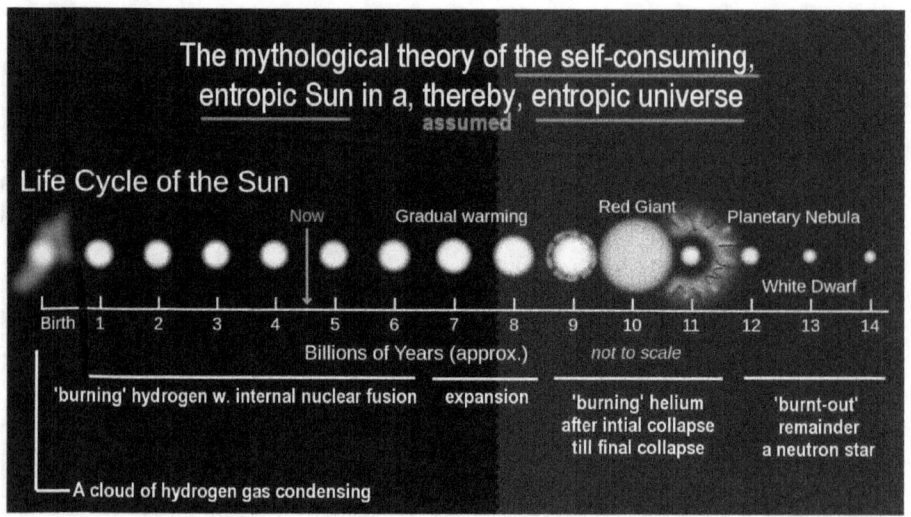

A sun is not entropic in nature. It is not self-consuming to produce energy. An energy-producing sun does not exist. The nuclear- fusion energy production that aims to replicate the energy production of the Sun is a mistake in premise.

The universe is energy

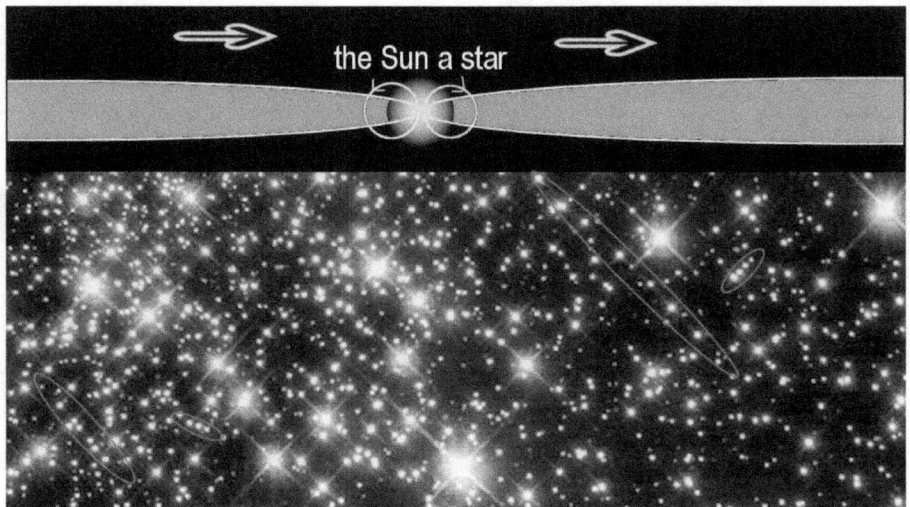

The universe is not consuming itself to produce energy. The universe is energy, and is producing everything with it.

We fail aiming for fusion power

We fail, when we aim to operate outside the platform of the universe, aiming for fusion power for which no basis exists. Indeed, we do fail big time in every single one of our attempts to do this.

Dead-end energy future

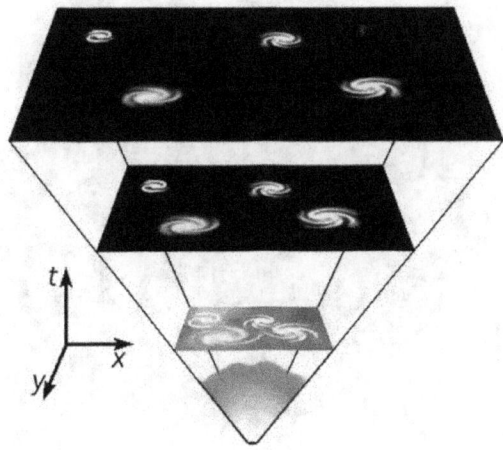

The point-source and dead-end energy future that the Big Bang theory parades as a concept, is false.

Explosion is entropic and blows itself out

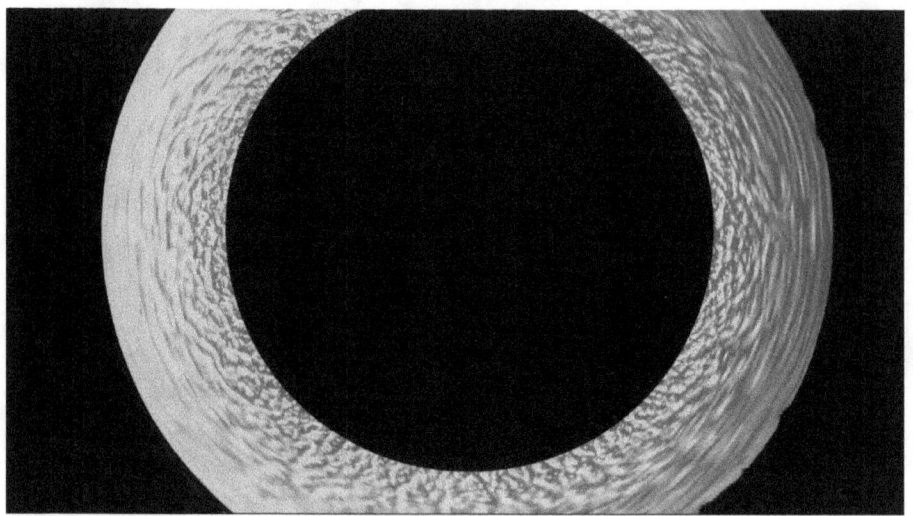

Every explosion is entropic and blows itself out. No evidence exists for it being the universal order. Every form of evidence reveals the theory to be a trap. We should celebrate that the theory is false, because if the Big Bang was real, humanity, would have no future.

By clinging to the Big Bang trap

strings of galaxies and stars

ESO/VIMOS galaxy cluster ACO 3341

ESA/Hubble & NASA.
Acknowledgement: Claude Cornen

At the present time, by clinging to the Big Bang trap, the trap of universal entropy, humanity denies itself to have a future. It denies itself the universal energy resource, which is a resource that is not based on a consumable fuel, but is active everywhere.

The anti-entropic energy resource

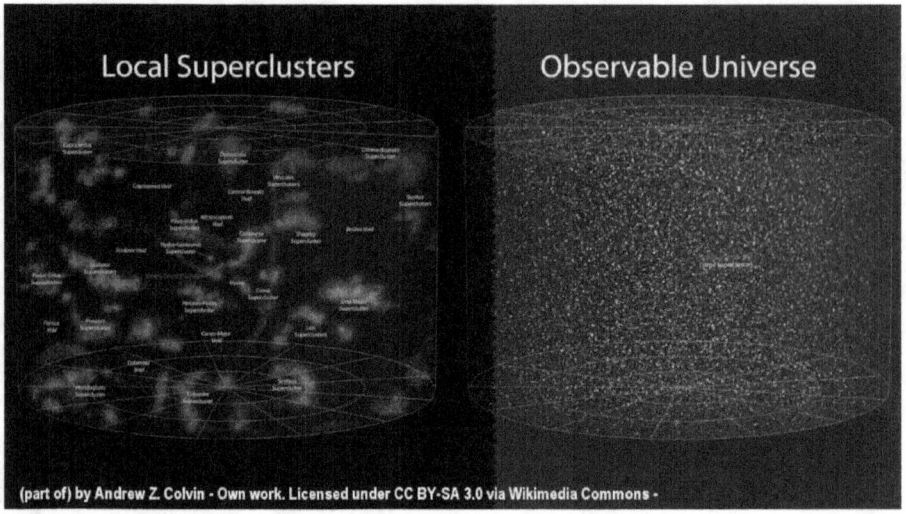

Local Superclusters   Observable Universe

(part of) by Andrew Z. Colvin - Own work. Licensed under CC BY-SA 3.0 via Wikimedia Commons -

The anti-entropic energy resource that is rooted in the nature of the universe itself, is a resource that is constantly self-renewing. The cosmic energy resource is plasma - plasma in motion - it powers everything. Our future depends on tapping into this resource. Without it we have no future. As I said before, all of our gas, oil, coal, nuclear energy systems, are fuel-based systems that are inherently entropic systems. These entropic energy systems are depleted by energy use. They are dead-end systems. Only the cosmic-energy system is self-renewing and self-expanding.

*If the Big Bang was the truth

The Crab Nebula

NASA

If the Big Bang was the truth, a nebula, such as the Crab nebula, would not exist. The crab nebula emits light energy at the equivalent of 75,000 Suns. Under the Big-Bang entropic theory, the light is the residual energy of a super-nova star explosion back in 1054 AD. However, no explosion, anywhere, has ever created a source of light that does not dim, but grows brighter. The light emitted by the Crab nebula, or any other nebula, is not residual energy from an explosion, but is light emitted by atoms activated with the movement of interstellar plasma that flows through the nebula, typically centered on a large sun.

Plasma flowing through a nebula

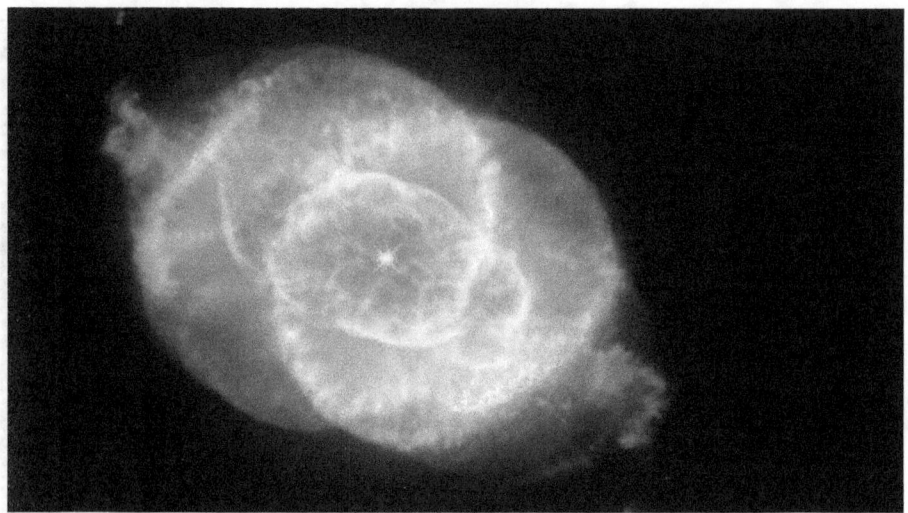

The flow-through process, of plasma flowing through a nebula, is indicated by the bipolar shape of nebulas that is often plainly visible.

Our energy-future on Earth

In these types of focused plasma streams, our energy-future on Earth is located. The Earth orbits within a highly concentrated plasma flow that presently creates the solar plasma environment.

The Red Square nebula

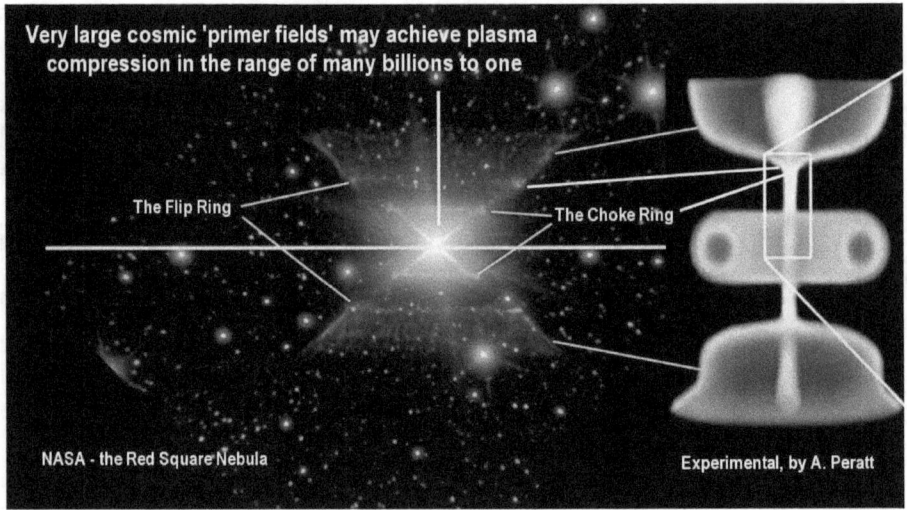

Very large cosmic 'primer fields' may achieve plasma compression in the range of many billions to one

The Flip Ring

The Choke Ring

NASA - the Red Square Nebula

Experimental, by A. Peratt

In some rare cases the shape of nebulas, such as in the case of the Red Square nebula, matches in principle the geometry observed in high-energy plasma experiments.

Our interface with the stellar plasma streams

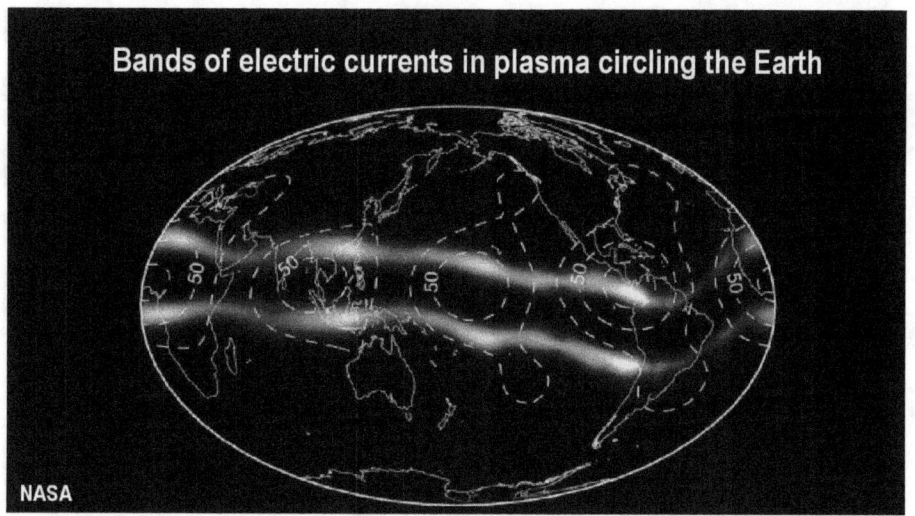

And as I said before, our interface with the stellar plasma streams, is visible in the form of equatorial plasma bands surrounding the Earth.

Also visible on the Sun

The same type of bands are also visible on the Sun that operates by the same electric principle, only much more powerfully so.

The cosmic electric energy platform

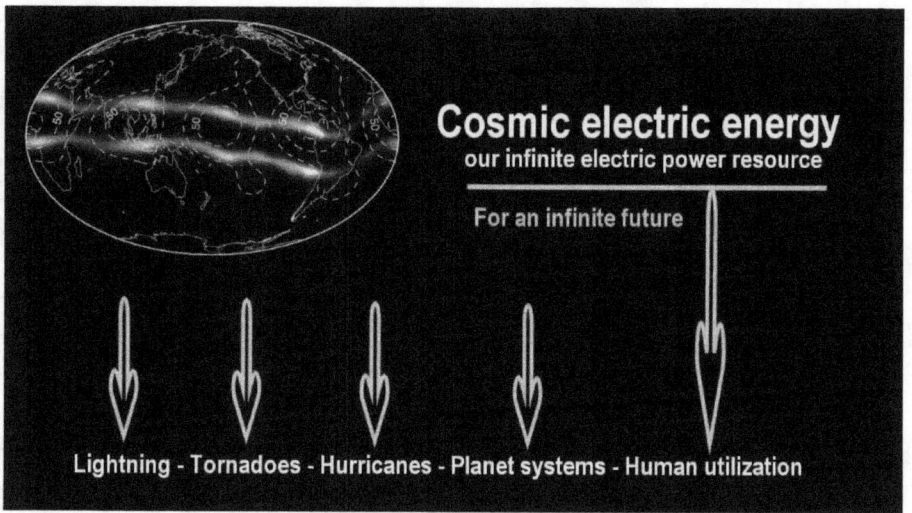

As I also mentioned before, the cosmic electric energy platform operates a number of large natural planetary systems, such as lightning, tornadoes, and the major global ocean and air currents. In comparison with the energy flux that powers these immensely powerful systems, our human energy needs are rather small.
It is this rather small additional cosmic energy utilization, that we will need to power our future with, which the Big Bang theory would deprive us of by its dream premise that plasma streams in cosmic space do not exist. By its premise, mass and gravity are the only forces that are recognized to exist as a basis for all effects.

The Big Bang Suicide Pact

"The Big Bang Suicide Pact"

The Big Bang theory is a trap

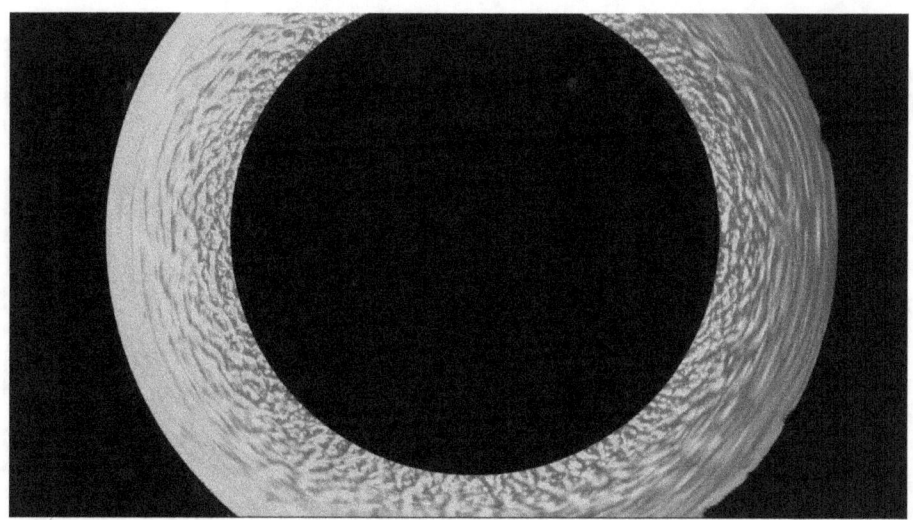

The Big Bang theory is a trap that is empty, a ring of smoke without substance. We always come back to this form as a model for entropy.
In the small-minded prison of universal entropy, the Big Bang theory acts as a global suicide contract that enforces energy starvation, and all kinds of related forms of starvation.

The biofuels genocide contract

The western imperial system is presently murdering 100 million people a year with the biofuels genocide contract that demands that vast quantities of high-value food are being burned as fuel in automobiles in a world that has a billion people living in chronic starvation.

The consequences of the difference

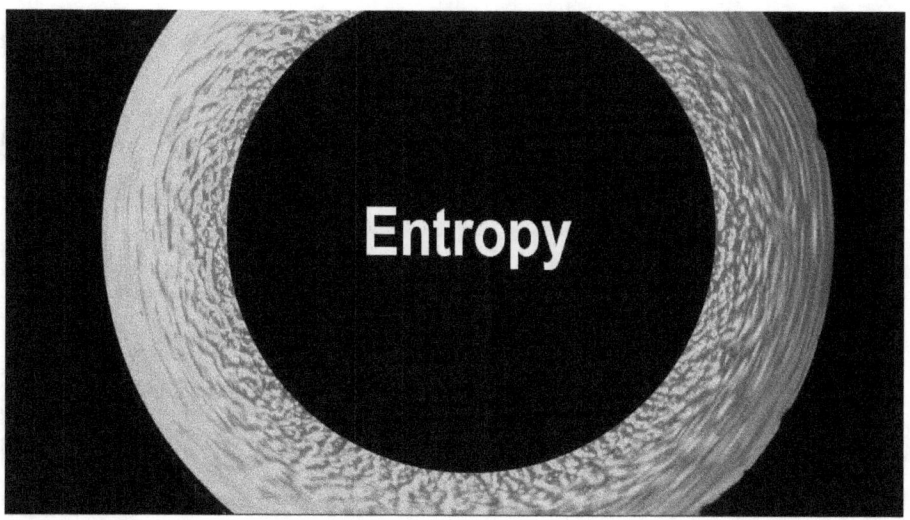

The consequences of the difference between the Big Bang cosmology, and the plasma cosmology that the Big Bang is designed to obscure, renders the controversy a deadly affair, which goes far beyond being merely an academic issue. The difference between the Big Bang system of entropy, and universal anti-entropy, is so vast that it has become an existential issue on the global scale. If the Big Bang Cosmology is not aborted, much of humanity will die in the near future from the imposed energy starvation and related insanities, all as the result of the assumed entropy of the system that society has been taught that it is bound to.

In this trap, the Sun is deemed entropic

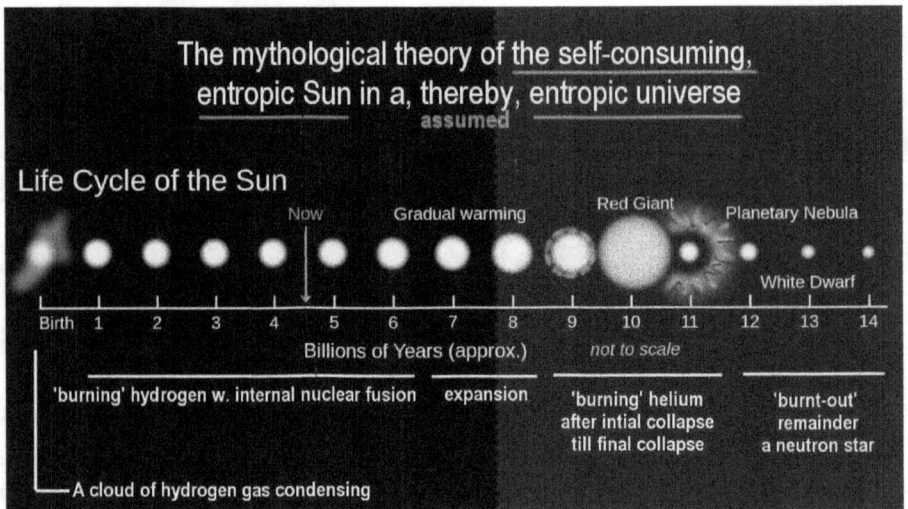

The mythological theory of the self-consuming, entropic Sun in a, thereby, entropic universe
assumed

Life Cycle of the Sun

Now   Gradual warming   Red Giant   Planetary Nebula

White Dwarf

Birth 1   2   3   4   5   6   7   8   9   10   11   12   13   14

Billions of Years (approx.)   *not to scale*

'burning' hydrogen w. internal nuclear fusion   expansion   'burning' helium after intial collapse till final collapse   'burnt-out' remainder a neutron star

A cloud of hydrogen gas condensing

Because the Big Bang universe is deemed to have evolved from gravitational condensation of primordial dust and gases into stars, planets, and galaxies, no energy resource is deemed to exist that maintains anything. In this trap, the Sun is seen standing by itself, deemed entropic and burning itself up, for the lack of recognition of the cosmic energy resource.

This dying-star solar model

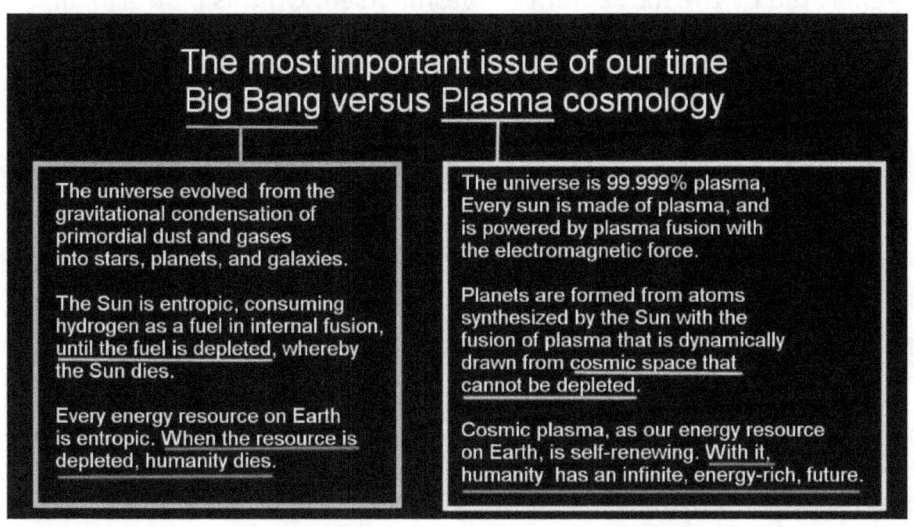

The most important issue of our time
Big Bang versus Plasma cosmology

The universe evolved from the gravitational condensation of primordial dust and gases into stars, planets, and galaxies.

The Sun is entropic, consuming hydrogen as a fuel in internal fusion, until the fuel is depleted, whereby the Sun dies.

Every energy resource on Earth is entropic. When the resource is depleted, humanity dies.

The universe is 99.999% plasma, Every sun is made of plasma, and is powered by plasma fusion with the electromagnetic force.

Planets are formed from atoms synthesized by the Sun with the fusion of plasma that is dynamically drawn from cosmic space that cannot be depleted.

Cosmic plasma, as our energy resource on Earth, is self-renewing. With it, humanity has an infinite, energy-rich, future.

In this trap, the Sun is assumed to be an isolated entity that is burning the hydrogen atoms that it is deemed to be made of, burning them as a fuel in internal nuclear fusion processes, until the fuel is depleted, whereby it dies in numerous ways. This dying-star solar model is then applied to the terrestrial energy resource model, which renders our future subject to depletion. Little hope remains after this. When the Earth's energy resources are depleted, and we are getting close to that, humanity has no option under this model but to die as a consequence, because no external energy resource is deemed to exist that would keep humanity going forever.

Now compare this inherent death spiral, with its opposite, the plasma cosmology. Here the universe is made up of plasma. This means that 99.999% of the universe, as plasma, carries an electric charge, and thereby electric energy. This means that a sun is not made up of cosmic dust and gases, but is made up of plasma with an electric quality, whereby it becomes electrically powered with a process of interstellar plasma becoming drawn to it that becomes fused into atoms on its surface with the electromagnetic force generated by the movement of plasma itself. On this platform all planets are formed from atoms synthesized by the Sun. Since plasma in cosmic space is self-renewing, it cannot be depleted. Consequently, a sun cannot die. Neither can the Earth ever be deprived of an energy resource, because the cosmic plasma that we can tap into, is not a fuel that can be drawn to depletion, but is an enduring quality of the universe. With it, humanity has an infinite, energy-rich, future.

For as long as the Big Bang cosmology rules

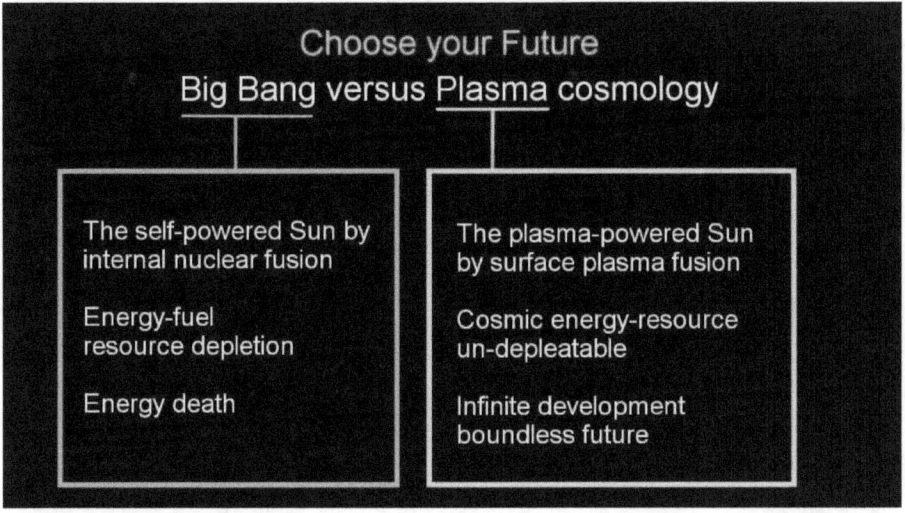

This means we have a choice before us. For as long as the Big Bang cosmology rules in the mind, humanity is effectively trapped by it, to die the death of the assumed energy resource depletion. This belief is apparently so strong that plans are promoted to harvest heliom-3 from the moon as a fusion-energy fuels, in spite of the fact that the last experiment required a million times greater energy input to cause the helion-3 fusion to happen, than it gave back as released energy. That's how heavily the assumption of energy entropy weighs on humanity, that it looks for miracles to circumvent it, while remaining unaware of the anti-entropic energy option that the Big Bang theory obscures.

The Big Bang entropic cosmology operates as this type of deadly package that includes numerous similar aspects.

Inversely, in the Plasma Universe, that stands as the total opposite of the entropic theory, everything is recognized to be actively powered by the forever-flowing cosmic, plasma energy streams.

The Plasma Universe, too, operates as a package. This package includes the plasma Sun, with cosmic plasma fusion occurring on its surface. It promises humanity an energy resource to tap into, that cannot be depleted, whereby humanity has an endless energy-rich future.

This leaves us with the question: which package will determine our future? The Big Bang package offers death by depletion of everything. The Plasma Universe package offers unlimited resources for development, life and abundance. Which of the two option would you choose as a basis for building a civilization on?

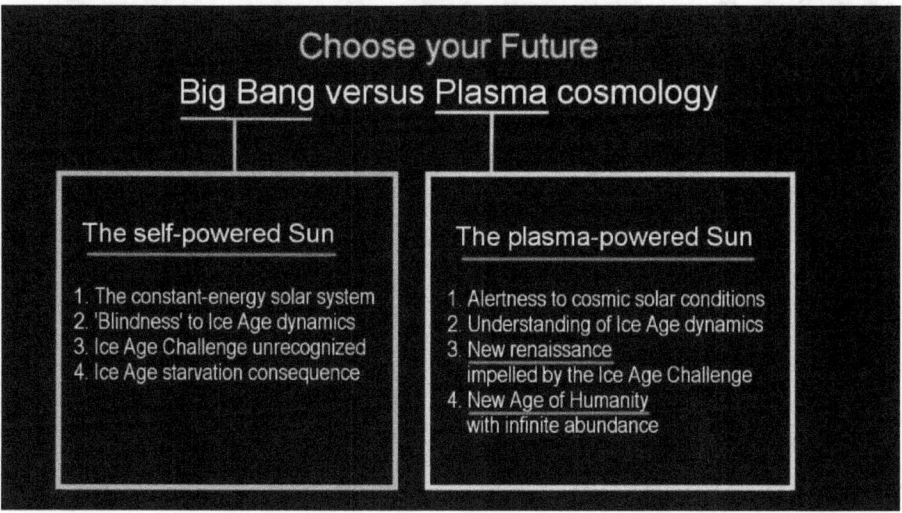

The Big Bang offers a trap. The trap has numerous faces. Each face reflects the notion of the self-powered Sun, whereby it hides the coming Ice Age and its consequences.

In contrast, the Plasma Universe offers itself as an open door. The open door has likewise many faces, with each reflecting the anti-entropic plasma-powered Sun and its electric principles that determine the Ice Age dynamics, which are knowable and enable us to respond to the dynamics to protect our civilization.

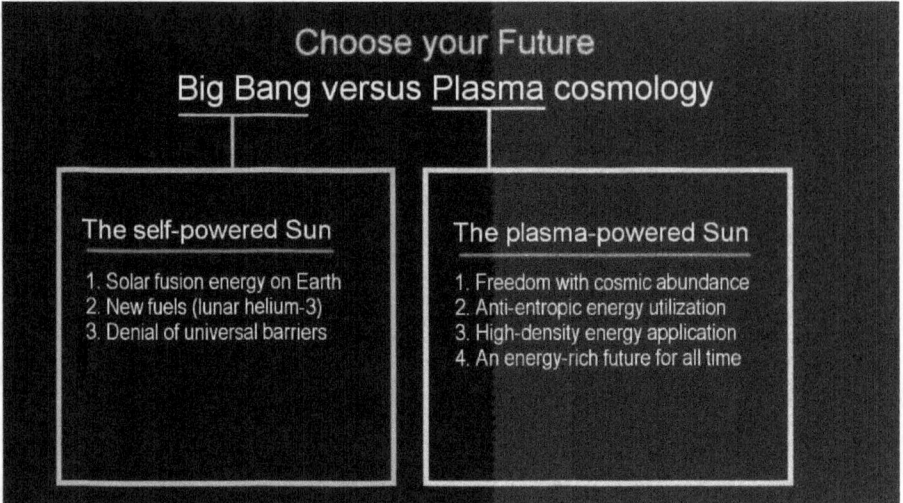

The Big Bang theory includes a number of entropic elements that set up traps in the mind that inspires desperate measures:

1. The solar fusion energy dream on Earth is a dead-end trap, because the Sun is not powered by internal atomic fusion that society attempts to replicate.

2. Since fusion power doesn't work, out of desperation, society now looks to the moon to mine helium-3 from its dust, in the desperate hope that helium-3 will be a useful fuel, though it is the hardest element to fuse.

3. The desperation causes a denial in science, of the universal barriers that the universe has erected against atomic fusion to protect its integrity.

The resulting entropy-inspired atomic-fusion energy hope stands nevertheless as but a brilliant dream with an empty center.

The Plasma Universe, in comparison, offers humanity real ready-made energy without the need for burning any type of fuel. It offers humanity

1. A great energy-freedom with cosmic abundance

2. It enables anti-entropic energy utilization
3. It enables high-density energy applications
4. It opens the door to an energy-rich future for all time to come
In the real universe, the plasma universe, the Earth is afloat in a sea that is energy, which powers the Sun. Why then would we need to 'produce' energy?